An Athenian Trireme Reconstructed

The British sea trials of *Olympias*, 1987

edited by
J. S. Morrison and J. F. Coates

BAR International Series 486
1989

B.A.R.

5, Centremead, Osney Mead, Oxford OX2 0DQ, England.

BAR -S486, 1989: 'An Athenian Trireme Reconstructed'

ISBN 9780860546238 paperback
ISBN 9781407347745 e-book
DOI https://doi.org/10.30861/9780860546238
A catalogue record for this book is available from the British Library
This book is available at www.barpublishing.com

OLYMPIAS

CONTENTS

LIST OF ILLUSTRATIONS

Frontispiece: *OLYMPIAS*

PREFACE

The Trireme Trust was founded in 1983 with the scientific and educational aim of building a reconstruction of an ancient Greek trieres, whose English name is trireme derived from the Latin triremis. The scientific purpose was to verify by experiment a theory of the trieres oarsystem and to discover through practical trials of a reconstructed ship the skills needed to use her under oar and sail, and in particular her performance and potentiality in combat and on voyage. The educational purpose was to create a ship which would illustrate in the most realistic way, not hitherto possible, the history of a period, crucial to the development of western civilisation, in which she formed the principal weapon of navies exercising economic and political power in the eastern Mediterranean.

The story of how this aim has been achieved has been told in two books (*The Athenian Trireme* Cambridge 1986 by J.S.Morrison and J.F.Coates and *Building the Trireme* London 1988 by Frank Welsh) and needs no repetition here. The two chapters introducing the report of the trials carried out by the British crew at Poros in 1987 summarise the ancient evidence on which the reconstruction has been based (Ch.1) and explain how this evidence was translated into an actual ship, now named *Olympias*, as the outcome of Anglo-Hellenic enterprise (Ch.2). The following chapters (3-5) are concerned with the actual trials of 1987 with notes, where relevant, to the trials of 1988, a full report of which will be published by an Anglo-Greek committee.

The Hellenic Navy and the Ministry for Cultural Affairs and Science generously put the reconstructed ship, under the command of Captain Papadas and with a Greek deck crew, at the disposal of the British Trireme Trust by whom the trials here described were organised. The Trust wishes to express its deep gratitude for their enthusiasm, generosity and imagination, and to thank Captain Papadas in particular, for making these trials such a rewarding and memorable event.

It should be emphasised, in the first place, that the actual building of the ship was entirely funded and managed in Greece, and that the ship is now state property in the charge of the Hellenic Navy. In addition to making the ship available to the Trust for the two weeks of the trials, the Hellenic Navy, by making accommodation available free of charge for the whole oar crew and trials staff at the naval petty officers school (S.M.Y.N.) at Poros, enabled the wholly voluntary oar crew, mainly students, to afford the cost of the fortnight's 'holiday'. Thus the trials could not have taken place without this generosity on the part of the Hellenic Navy, nor could the virtually simultaneous transport of about 200 people by air have been arranged without the administrative assistance of the Ministry of Cultural Affairs, and in particular of John Skotiniades acting on the Ministry's behalf. The Trust acknowledges and wishes to express its gratitude for this crucial assistance as well as for the support for this Anglo-Hellenic enterprise given by the British Embassy, the Hellenic Maritime Museum and among the many Greek and British individuals whose help and interest has done so much to carry the Trireme Project along, in particular Mr George Dracopoulos and Mr Harry Tzalas.

The Trust wishes to express its great gratitude to our colleague Mrs Rosie Randolph who with enormous energy and patience organised the recruitment of the oarcrew and the transport and accommodation of them and those who accompanied them as well as a number of British visitors; also to John Illsley, who as official photographer for the Trust made a complete and detailed record of the trials with great skill and efficiency.

The Trust also wishes to express its gratitude, and warm congratulations, to the trials staff and oarcrew who came out at their own expense to man and row a (to them) formidably strange and novel craft less than a month after her completion. The success of the trials is mainly due to their marvellous enthusiasm and spirit, maintained in the heat of the summer of 1987, great even by Greek standards.

The editors are grateful to Mr Christopher Dodd, the editor of *Regatta*, for permission to reprint from the issue of November 1987 the article 'the Rowing Master's Story' by Michael Budd (Annex B 9 p.108-9).

They also wish to thank Professor Sean McGrail for advice on the form of this report and for his helpful detailed comments on the draft text.

Readers are referred to the forthcoming *Proceedings of the Second Symposium on Ship Construction in Antiquity (Delphi August 1987)* in which a paper will appear by Mr Owain Roberts, Sailing Master during the Sea Trials at Poros, on the design of the rig of *Olympias*.

On behalf of the Trireme Trust:

Editors: J.S.Morrison and J.F.Coates

Chief Recorder: I.M.Whitehead

ABBREVIATIONS

AT See Bibliography

CQ *Classical Quarterly*

F Fahrenheit

GOS See Bibliography

Hdt. Herodotus

ibidem in the work previously cited

IG See Bibliography

IJNA *International Journal of Nautical Archaeology*

JHS *Journal of Hellenic Studies*

kg. kilogram

kn. knot

kph kilometres per hour

Meiggs-Lewis See Bibliography

MIMA See Bibliography

S.M.Y.N. The Petty Officers' School at Poros

s.p.m strokes per minute

Thuc. Thucydides

Xen. Xenophon

See also p.77 for abbreviations used in Annex B 4

CHAPTER ONE : THE EVIDENCE

THE EVIDENCE : John Morrison

1.1.INTRODUCTION.

The sea trials of a reconstructed Athenian trieres which are recorded in this report are the culmination of much work by many people on the history and development of the Greek oared warship. This work has been bedevilled ever since the Renaissance by an apparently endless controversy about the nature of the trieres' oarsystem.

In 1968 R.T.Williams and I produced in *Greek Oared Ships* (= *GOS*) an assembly of all the then known evidence for the development of these ships in the period 900-322 BC, in which of course the trieres figured largely. Up to 1968, as a result of the technical advances in diving gear during the second World War, underwater archaeologists had discovered a number of ancient wrecks on the seabed of the Mediterranean, but none of warships.The evidence for the trieres set out in *GOS* was accordingly entirely secondary, i.e. not derived from the remains of an actual trieres. It has since become clear that this negative evidence indicated something positive, which had already been pointed out but little noticed, viz., that oared warships, carrying no ballast or cargo, had a positive buoyancy and when holed were swamped but did not sink. Phoenician oared ships with protruding forefeet have been found on the seabed off Marsala in Sicily, but the ballast which they carried makes it probable that they were oared merchantmen of a type known from representations, presumably armed with a ram to deter pirates or enemies.

When in 1981 the plan to build a reconstructed Athenian trieres began to take shape, two things were realised: first that it was unlikely that remains of a trieres would be found on the seabed, and secondly that even if some were found the experience with merchantmen indicated that only the lower timbers of the hull would be preserved and these would give no clues to the oarsystem upon which the reconstruction would largely have to be based. The conclusion was that for the trieres in 1981 as in 1968 the evidence would have to be secondary. This evidence and the design based upon it has been set out at length by John Coates and myself in *The Athenian Trireme* (=*AT*) and will be summarised here as methodically as possible. It falls under four heads, placed in order of scientific respectability: archaeology,land-surface and submarine; epigraphy; literature; and unlabelled iconography. Of this some is direct, i.e. relating specifically to triereis, and some is indirect, i.e. relating to triereis only by inference.

1.2 ARCHAEOLOGY

Underwater archaeology has now shown that the universal method of hull construction in the Mediterranean in antiquity roughly since the beginning of the Bronze Age was what is known as a shell method, by which the garboard strakes are fastened to the keel on each side, and the other planks thereafter in succession, laid edge to edge, by mortise-and-tenon joints every few inches. The evidence comes from merchant ships of all periods, but the find of a bronze ram in the sea off Haifa, still containing some timbers, probably of a large Hellenistic warship, confirms what was already highly likely, that the shell method was also used in warship construction. If the hull of the trieres was built by the shell method, it must necessarily have the characteristic 'wine-glass' shape, found in every wreck of the period including the long, oared, ships excavated off Marsala.

A third type of archaeological evidence is provided by the remains of the ship-sheds in which triereis were normally kept. In 1886 Dragatzes published the results of an excavation the previous year at Zea in Piraeus which had revealed the remains of the ship-sheds rebuilt in the fourth century BC on the rock-cut foundations of the sheds destroyed after Athens' defeat in 404 BC. Between each shed and the next was a row of pillars, in one case a solid wall, determining the maximum breadth of the ship. Each trieres rested on a stone slip 3m. wide running down to the water's edge at a gradient of one in ten. These slips were measured as having an average dry length of 37 metres. In no case was the distance to which the lower end extended into the water measured at the time of the excavation, although Dragatzes speaks of 'a considerable distance'. The failure is unfortunate. It is most unlikely that the ship would not have been pulled up right out of the water, but a change in the water level cannot be ruled out, although none is recorded. Nevertheless, 37 metres can be taken as a rough maximum for a fifth and fourth century trieres. The actual length will rest, to a large extent, on the building design of the ship, determined by a number of other factors of which the most important is the longitudinal space occupied by each oarsman, the 'room'(Glossary). This traditional unit, *dipechiake* in Greek, *interscalmium* in Latin, is said by Vitruvius (1.2.4) to be of two cubits (.888 m), the meaning of the Greek name. The length of an Attic cubit (1½ feet) is taken from measurement of the older temple of Athena on the Acropolis which was known as Hecatompedon (i.e. a hundred feet long).

1.3. EPIGRAPHY

The numerous inscriptions found in Piraeus and directly relating to the trieres may be divided into three categories :-

1.3.1. The earliest (IG.1³ 89 31 440-425 BC) is part of a treaty between Perdiccas, king of Macedon, and Athens, in which the king undertook to export to Athens oar timber and to Athens alone. This timber (*kopeus*) appears from Theophrastus 5.1.6-7 (see *AT p.190*) to have consisted of young fir trees of suitable height and girth. A second inscription, from about the same period, is from a decree laying down that a trieres should be manhandled up the slipway of a ship-shed by not less than [one hundred and] forty men, and down by not less than [one hundred and] twenty men. The same inscription lays down that not less than [fifty] men should be employed to fit the *hypozoma* (see below). The numbers of men specified in each case (according to D.M.Lewis's very probable guesses) indicate that for the safety of the ship an operation on a considerable scale was required.

The weight of the *hypozoma* is given in an inscription dated after 314 BC. (IG.2² 1479 B 49 cf.57: *GOS p.295-6*) and its length may accordingly be estimated roughly at about 90m or 300 ft. At this date the rope could belong equally well to a tetreres or penteres as to a trieres. In any case it is clearly long enough to be fitted, doubled, from bow to stern in any of these warships.

It is interesting that the Alexandrian epic poet, Apollonius of Rhodes, links together, as does the Athenian decree, the fitting of the *hypozoma* with the launching of a ship, in his case the legendary Argo. He says

(I 363-393):' First of all they' (Jason and all fifty heroes, the crew of the pentecontor)' girded the ship mightily with a well twisted rope from within, putting a tension on each end so that the planks should fit well with the tenons and should withstand the opposing force of the sea's surge'. Both the method of fitting and the purpose of the *hypozoma* is clearly stated by the poet.

Next, after digging a trench from the place where she had been built down to the sea and laying wooden sleepers in it,'they tilted the ship down on to the sleepers so that she might run over them and be slipped. Then high on both sides of the ship they turned up the oars and tied them about the tholepins so as to project a cubit's length (.444 m or 17½ inches). In single file on each side of the ship the heroes stood and pushed with chest and hands (viz. against the projecting oars). Tiphys, the *keleustes* or boatswain, went on board to urge the young men to move the ship in the proper manner. He shouted his commands mightily. Argo responded quickly....and slipped into the sea. But the heroes reined her back as she moved forward and held her in control'. Again, the poet makes it clear how a big oared ship was slipped and why a large number of men was needed for the operation.

Apollonius is likely to have derived his descriptions from his own observation of the preparation and launching of the warships, triereis and others, commissioned in the third century BC by his native city, a maritime power in the eastern Mediterranean. It may then be taken as an illustration both of the fitting of the *hypozoma* and the slipping of a trieres at Piraeus in the fourth century with a large body of men. A similar procedure, but with a somewhat bigger workforce, is likely to have been used in hauling a trieres up the one-in-ten gradient of a Zea ship-shed. In slipping and hauling up the use of numbers of men is obviously necessary. In fitting the *hypozoma* the large number of men was apparently needed to make it as tight as possible. *Tonoi*, tightening devices, are mentioned in connection with the *hypozoma* in inscriptions (IG 2^2 1610.23,1613 280-2,1673,12) as well as by Plato (*Laws 945 C*).

A further relevant inscription in this group, from the time of the Peloponnesian war(IG $1^3$199), contains the first reference to a type of trieres which frequently appears in the historians, the *stratiotis*, designed, or possibly adapted, for the transport of troops (see *AT p.156-7*).

1.3.2. The second category consists of a single inscription, IG.2^2 1951, on nine much damaged fragments of marble and dated at the beginning of the fourth century. The text appears to have contained the names of crewmembers of ships which at this date could only have been triereis. One list, more fully preserved than the other two, begins with two trierarchs (lines 79-81 cf.312-3 and 406-8). Then follow eleven *epibatai* (hoplite soldiers, lines 82-3 cf.315-25 and 409), a *keleustes* (boatswain, lines 96-7 cf.332-3), a pentecontarch (purser, lines 98-9 cf.336-7 and see *AT* p.124), an *auletes* (pipeman, lines 100-1 cf.334-5?), a *naupegos* (ship's carpenter, lines 102-3), a *prorates* (bow-officer, lines 104-5 cf.330-1) and three *toxotai* (archers,lines 106-109,or two, lines 185-7). The collective term for these soldiers and petty officers appears to have been *hyperesia*.

There are also the names of *nautai astoi* (citizen seamen,lines 110-16 cf.188ff) followed by the names of a number of *therapontes*(slaves,lines 117-84). This latter entry has suggested that the lists may be of casualties in the battle of Arginusae where slaves served in an emergency fleet with the promise of freedom. But the presence in them of two trierarchs makes this an unlikely possibility since although two trierarchs was regular practice at this time for financial reasons both would hardly have been on board at one time.

1.3.3. The third group of inscriptions relating to the trieres is the most important. These, known as the naval inventories, consist of a series of lists of the ships and gear in the naval dockyards at Piraeus. Incompletely and intermittently they cover the period 377/6-323-2 BC. The information they give falls under the following headings;-

1.3.3.1. *Number of Ship-sheds*

IG $2^2$1628 (326/5) 481,553 : 372 ship-sheds of which 82 were in Munichia, 195 in Zea and 94 in Kantharos. In IG 2^2 1611 6 (357/6) some of the triereis listed are *hupaithrioi*, not under cover.

1.3.3.2. *Number of Ships*

In the surviving fragments of the earlier naval inventories the total number of ships in the dockyards is not given. IG 2^2 1611 1-9 (357/6) records a total of 283 ships in the dockyard, and IG 2^2 1613 284-302 (353/2) a total of 349 triereis in the dockyard and at sea. IG 2^2 1628 481,553 gives 360 triereis and a number (lost) of tetrereis as well as the number of ship-sheds.

1.3.3.3. *Categories of triereis*

IG 2^2 1611 65-128 lists a category of triereis, mostly new, which are called *exairetoi*, select, and are presumably the same as the special reserve of triereis mentioned by Thucydides (2.24.2) and given the same name by him. Elsewhere in the same inventory triereis are recorded as in the first, second or third class, or merely as old. These are presumably the ratings of the ships not in the reserve. Thucydides says that old, but presumably still serviceable, triereis were first used as horse transports for the cavalry in 430 BC (2.56.2). They would have needed considerable adaptation (see *AT p.226-8*). IG 2^2 1628 154 ff speaks of such ships as pulled by sixty oarsmen.

1.3.3.4. *Themistocles' Decree*

There is a further document known as Themistocles' decree (Meiggs-Lewis 23, see *AT p.108 ff.*) which is thought not to record the actual decree proposed by Themistocles and passed by the Athenian Assembly in the autumn before the Salamis campaign, but a literary version current in the fourth century. It is nevertheless an important source of information about the Athenian navy, setting out the measures taken, or credibly believed in the fourth century to have been taken, by the Athenian Assembly in late 481 to meet the invasion which they believed imminent. These measures include the mobilisation of the fleet and its immediate deployment, and reflect uncertainty as to when the enemy was to be expected.

The measures are as follows:- The military commanders are to appoint by lot a suitably qualified trierarch for each of the 200 ships. They are to call up ten hoplites (armed infantrymen) aged between twenty and thirty,and four archers for each ship. The body of support staff for each ship, the *huperesia**, which included the military personnel mentioned, six petty officers, and two gangs of deck hands, thirty men in all, was also to be allocated to the triereis by lot, as the trierarchs had been. On the city notice board are to be posted by name the *huperesiai* as well as the *taxeis*, the crews of 170 oarsmen, for the ships. The two hundred ships thus manned are to be divided into two fleets of 100 ships each, one to be sent to the northern tip of Euboea to await the arrival of the Persian fleet if it came that way, and another to patrol Attica in case the enemy made a direct attack, as they had in 480.

1.3.3.5. *Ships' Gear* (all references in IG 2[2])

In the absence of more than a handful of representations which may be taken to depict triereis, none of them a complete vessel and none of them under sail, the lists of gear labelled as belonging to triereis of the fourth century are of particular value. Here the main items only, those which affect the structure and working of the ship, are noticed.

The picture of the fourth century Athenian trieres which the naval inventories enable us to draw is one of much useful detail, from the pair of rudders in the stern through the 'engine room' of 170 oarsmen to the bronze ram in the bows. Of the three categories of oars, and hence of oarsmen, we know that in the middle fourth century two consisted of 27 each on each side of the ship and one of 31. From the name *thalamios* it can be deduced that these oarsmen sat in the *thalamos* or hold, and from the name *zygios* that these sat on the thwarts, *zyga*, as Homeric oarsmen did in their pentecontors with twenty-five oarsmen a side at one level. The *zygioi* accordingly sat at a higher level than the *thalamioi*. From a passage in Aristophanes *Frogs 1704* it is plain that the *thalamioi* sat at a lower level than some at any rate of the other oarsmen. Since each of the oarsmen has his own oar (confirmed by Thucydides 3.93.2) and all the oars in one vertical triad (see Glossary) are of the same length, the thranite oarsmen must sit outboard and slightly higher than the *zygioi*. The evidence of literature and, to some extent of iconography, indicates that the thranite oars were worked through an outrigger, as their position necessitated.

The working oars are described collectively as *tarros*, the spare oars as *perineo* (cf.Thucydides 1.10.4 and e.g. 1610 (370 BC) 30): ' For this ship the *tarros* is complete, the *perineo* are missing'.The working oars are given in detail in a number of entries as 62 thranite, 54 zygian and 54 thalamian (1606 (374/3 BC), 1607 373/2 BC). There is reason to believe (*AT p.147*)that the numbers of oars in the three categories were 60, 58, 52 in the fifth century and that those numbers also prevailed in 377/6. The number of spare oars recorded is thirty. That there was a difference in appearance at any rate between the oars of the thranite and zygian categories is shown by an entry in 1604 (377/6 BC): the inspector notes that ' four of the thranite oars so listed are zygian '. The length of the working oars is not given but on one occasion in 1606 and on seven occasions in 1607 the length of the spare oars is given, viz. nine and nine and a half cubits (4.0 and 4.2 m).

These latter entries imply that the working oars, for which the spare oars were taken on board as replacements, must have been of these two lengths, differing by 22 cm.. The disparity,at first sight puzzling, is explained by Aristotle (384-322 BC) in his treatise *On the Parts of Animals* (687 b 18), where he says that in the human hand ' the end fingers are short rightly and the middle finger longer, as is the oar amidships......'. The Greek medical writer Galen of Pergamum, who lived and practised in Rome in the last three quarters of the second century BC. expands Aristotle's explanation very usefully (*On the Use of Parts 1,24*): ' As I think in triereis too the ends of the oars extend an equal distance (i.e. the part of the oar outboard of the tholepin is the same length throughout the ship) although they are not all the same length, because in fact in triereis too they make the midships oars the longest for the same reason'. Galen, like Aristotle, is explaining why the middle finger of the hand is longer than those on each side. Galen indicates a valuable point, that the oars in the bow and stern are shorter because the looms (the inboard parts) are shorter. This is also a feature of the oars of a modern whaler and for the same reason : the narrowing of the hull at each end does not leave room for the looms to be of full length.

The *hypozoma* has already been noted (pp.4-6) in connection with an early inscription in this group. When the *hypozoma* had been fitted the ship was ready for service (1627 (330/329 BC) 29). Four were the normal number carried, two more being added when there was the prospect of an abnormally long absence from Piraeus (1629 (325/4 BC) 11f.). They were each a good deal more than twice the length of the ship. The fact that they were issued in pairs, as well as other considerations, suggests that they were fitted and tightened one pair at a time. As was seen, the rope needed to be tight, and would thus have required constant tightening and before very long replacement. The entry 'pieces of *hypozomata*' in 1610 suggests that they broke in service.

Pairs of rudders are recorded (1606 (374/3 BC) 34,1608 (373/2 BC) 24 and 1611 (357/6 BC) 23-4) and of one rudder it is noted that its *pteryx*, wing, is broken..

A secondary means of propulsion is also attested with the provision of masts and sails and of a variety of rope tackle. Two masts are regularly listed, a main mast and a boat mast (*akateios*), similarly main sails and boat sails (1604 (377/6BC) *passim*. Two sorts of sails, thin/light and thick/heavy occur in three inscriptions after 330 BC (1627 665,1629 371, 1631 415-17). The absence of yards (*keraiai*) from the lists of wooden gear is surprising and difficult to explain. Possibly they were regarded as inseparable from the masts, or, more probably, from the sails, as in Homer.

A further detail of the oarsystem given by the inventories is the fitting of *askomata* (leather sleeves) to some of the oarports. Since the purpose of such sleeves was clearly to prevent the entry of water either in rough weather or when the ship heeled under sail, the ports so fitted must have been those of the thalamian oars.

The importance of *askomata* is shown by the fact that in the earliest naval inventory to survive, 1604 (377/6 BC), it is noted against each ship in the list either that the *askomata* are already fitted or that someone named has them in his possession, or that the dockyard has the money for them, or that they are missing. Their importance is confirmed in Aristophanes *Frogs 364* where they are included with flax and pitch as contraband articles whose export was forbidden in time of war. The value of a trieres' *askomata* is given as 43 drachmas 2 obols in the inventory abovementioned, five drachmas apiece if the number of thalamians was 52 as is possibly the case in the fifth century, and might have been the case in 377/6 BC.

There are numerous entries of screens, of two kinds, canvas and hair, for defence against the weather and enemy missiles (e.g.1611 244-249), and one entry of screens of leather, more efficient for the latter purpose (1627 348). Other standard equipment are ladders and anchors, two for each ship (e.g.1611 28 and 259). The bronze ram is a structural feature rather than ship's gear, but it occurs nevertheless in the inventories from time to time. If a ship was abandoned as unserviceable abroad, the ram was returnable to the dockyard by the trierarch and the debt is recorded (1623 (334/3 BC) 11-12,1628 (326/5 BC)).

1.4. LITERATURE

1.4.1. Archaeology and epigraphy have been seen to supply an amount of factual knowledge relating to the trieres, either directly or indirectly, which may now be elaborated by a few facts implied in literary passages.

In the comedies of Aristophanes performed during the Peloponnesian war (430-404 BC) there are many revealing passages showing in particular : the pride of the ordinary Athenian in his city's 'fine triereis'(*Birds 108*), his keenness to excel in pulling an oar (*Wasps 1097*), the lower position of the thalamians(*Frogs 1074*),and the importance to the city of the thranite oarsmen (*Acharnians 162-3*) .

A description of a scene in the dockyards when preparations, given in some detail, are under way for the dispatch of a naval expedition (*Acharnians 552 ff.*) and imaginary conversations between personified triereis (*Knights 1300-10*) and of horses about going on board cavalry transports (*ibidem 595-610*) are all informative in different ways.

From the point of view of the oarsystem the most important pieces of evidence given by the historical texts are provided by Herodotus and Thucydides.

From Herodotus' *History* we learn that in the sixth and fifth centuries the trieres had a complement of 200 men (3.13.1-3, 7.184.1,and 8.17.) in addition to the trierarch. This number seems to have remained constant throughout the 4th century (*AT* p.108). Another piece of information derives casually from the story of Scylax, the Milesian trierarch in the fleet of 200 triereis financed by Persia for the conquest of Naxos in 500 BC. As punishment for failure to keep a proper look-out on his ship, he was tied up with his head through a thalamian oarport. This story (5.33.2) tells three things, that triereis had oarports, that thalamian oarports were not the only oarports, and thirdly that the oarport specified was big enough for a man's head to pass through.

The two passages from Thucydides' *History of the Peloponnesian War* are treated at length in *AT pp.163 and 166*. They both concern the part of the trieres called the *parexeiresia*. The word, which means ' a rowing outside along ', suggests an outrigger. If there were, as it appears, more than one oar worked through oarports, three levels of oars in all and an outrigger, it is reasonable to conclude that the oars of the lower two levels were rowed through the oarports and the highest through the outrigger. In the first passage (4.11.4 -12.1) Brasidas,attempting to land from the deck of his ship by means of a ladder, was wounded and lost consciousness, falling into the *parexeiresia*. The second, more precisely informative, passage(7.34) concerns an engagement between an Athenian squadron of thirty ships based at Naupactus on the northern shore of the Gulf of Corinth and a Corinthian force of about the same number. The latter formed up in line abreast in the bay of Erineus with each wing protected by a headland, thus preventing the faster Athenian squadron from outflanking them with a *periplous*. In the battle which ensued seven of the Athenian ships were put out of action having been rammed bow to bow and their *parexeiresiai* damaged by the *epotides* of the enemy ships which had been specially strengthened for this purpose.

CHAPTER ONE : THE EVIDENCE

Theophrastus remarks (*Enquiry into Plants* 4.2.8.) that the *epotides* of a trieres, like the stempost, were made of ash, mulberry, or elm since these parts of the ship had to be tough. The word *epotis* is descriptive , *epi* = on and *ous, otos* = ear, a heavy beam laid across the bows of the ship, the projecting ends of which resembled ears. The bows of ships depicted on black-figure attic vases are often shown in the form of an animal's head. When ships rammed each other head-on (a tactic avoided if possible), the rams would slide past each other, and the impact would fall on either the port or starboard *epotis* of both ships.

Since the Corinthians had specially strengthened the *epotides* of their ships with this tactical situation in mind, and contrivred a battle in which it was unavoidable, fatal damage to the Athenian ships followed, since they were more lightly built, for quick acceleration and sharp turns.

On Corinthian advice the Syracusans, besieged at this time by an Athenian naval force which they had succeeded in confining to the Great Harbour where superior speed and agility were useless, also strengthened the bows of their ships 'laying stout *epotides* thereon and fixing stays from these *epotides* to their ships' sides inside and out'. In the Great Harbour at Syracuse as in the bay of Erineus the Athenians were denied the opportunity of *diekplous* and *periplous* in which their skill lay. These passages show very clearly the position of the *epotides* in the upper part of the ship and consequently of the *parexeiresiai* which they protected.

Two passages will be sufficient to show the presence of a canopy deck, the *katastroma*, in fifth and fourth century triereis (*AT p.159-162*). The historian Xenophon, describing (*Hellenica 1.4.18*) the return of Alcibiades to Piraeus in 407 BC, says that he did not immediately disembark but went up on deck to see if any of his friends were there to welcome him; and Herodotus tells the story (8.118.1-4) of the ship in which Xerxes was returning to Asia Minor after Salamis. She was overloaded, the deck packed with a large number of Persian notables of his entourage, enough to endanger the safety of the ship. The ships in which Alcibiades and Xerxes were travelling would certainly have been triereis.

1.4.2. The principal and most important contribution of the ancient writers is to present a picture of the sort of ship the trieres was in general and what tactical and strategical aims she was designed and built to achieve. The reconstructed ship, besides satisfying the physical requirements of the ancient evidence, must be such as to match this picture, with a properly trained and experienced crew, in performance, combat manoeuvres and in deployment.

1.4.2.1.*Combat*

The tragedian Aeschylus and the historians Thucydides and Xenophon give accounts of naval actions in which Athenian triereis were engaged during their own lifetime. Aeschylus is said to have taken part in the battle of Salamis, which he describes in his play, the *Persians*. Thucydides commanded a force of triereis in the war which was the subject of his history and lived to see its conclusion. Xenophon was a historian and a writer on many subjects who was also a soldier, and he must have seen triereis in action. His description of the battle of Arginusae in 406 is particularly detailed. Herodotus, who wrote about the Persian wars in the middle quarters of the fifth century BC, was a native of the Greco-Carian city of Halicarnassus (mod.Bodrum), which contributed a small squadron of triereis to the Persian invasion fleet in 490. His information about the tactics of fighting with triereis is interesting and likely to be based on first hand knowledge. These accounts by writers whose credentials are excellent are treated in detail in *AT Chapters 3-5*. The only complaint a modern reader can make is that they assume a knowledge of much that would have been commonplace to their contemporaries, but must now be for him a matter of inference and extrapolation. Here the conclusions are summarised.

Herodotus displays a special interest in the tactics of fighting at sea with the ram, mentioning an early battle in which pentecontors had their rams 'twisted off', and the training sessions of the Ionian fleet in revolt against Persia in 500 BC (1.166.2 and 6.11.2-15.2). In the latter the elected commander, Dionysius of Phocaea,' took ships to sea each time (i.e. squadron by squadron) in line-ahead formation so as to give the oarsmen work making a *diekplous* with the ships through each other (i.e. one squadron through another), and to have the decksoldiers under arms. His description is elliptical, but if ships were in line-ahead formation in preparation for the *diekplous*, the operation must have been one not of single ships but of a squadron in line ahead attempting to penetrate the opposing squadron, which was in the normal defensive formation of line abreast. And this 'breakthrough', if it entailed arming the decksoldiers, envisaged, or at any rate risked, the possibility (in a real action) of close combat involving the decksoldiers.

Aeschylus's words in his account of the battle of Salamis make it quite clear that the main armament of the trieres was the ram, not the decksoldiers (lines 336, 408-9, and 415). The initial formation of fleets in a naval confrontation was normally, but not always, line abreast (*metopedon*). Exceptionally, e.g. at Arginusae, the faster fleet came into battle in line ahead. There, Xenophon says (1.6.29), the Spartan fleet rowed into action 'the ships all in single file ready for the *diekplous* and *periplous* because they were the faster fleet'. At Salamis the mainland Greek fleet, though the slower, in Aeschylus's account took the initiative against ships which though technically faster had been at sea under oar all night, moved from line abreast to line ahead and attacked the left wing of the Persian line, broke through after close combat and encircled the Persian ships.

In the later fifth century it is plain that speed and lightness were the characteristics which Athenian triereis were designed to possess (Xenophon *Household Management* 8.8, Theophrastus *Enquiry into Plants* 5.7.4.); so that the manoeuvres of the *diekplous* and *periplous* could be neatly carried out, thus removing the advantage of greater numbers and bringing the ram to bear in the most effective manner. Thucydides, in his account of the battle of Sybota in 432 (1.45-54: see *AT p.64*), speaks of the 'old-fashioned method of fighting at sea, more like a land battle with ships immobilised and no opportunity for a *diekplous*', the tactic of swift movement. Slow ships preferred confined water, to avoid encirclement, fast ships open water in which they could encircle the enemy. Phormio, tactically the Athenian Nelson, observes (Thucydides 2.88) : ' in a confined space 'the fact is that one cannot properly move against a ship to ram it, if one cannot get the enemy in one's sights from some way off.There are no opportunities for a *diekplous* or a sharp turn which are the manoeuvres of a faster moving fleet .... and numbers are decisive'.

To counter the lightly built Athenian ships, accustomed to ram *ek periplou*, i.e.,after encirclement, in the stern or quarter, and not bow-to-bow, we have seen that the Corinthians in the Gulf and, following their example, the Syracusans in the Great Harbour once they had confined the

Athenian fleet there, strengthened the bows of their ships, and so denied the Athenians the 'breakthrough', while the confined water denied them the alternative manoeuvre, the 'encirclement'.

From the battle situations described very briefly it is clear that good acceleration and a good turning capacity, to achieve a strike most effectively, as well as the ability to withdraw quickly by backing water, were the qualities which Athenian commanders in the fifth and fourth centuries looked for in their triereis and which ancient crews were capable of achieving. These records provide the designer of a modern reproduction, and the crews that man her, targets at which to aim.

1.4.2.2.*Deployment*

The trieres was designed primarily to destroy other warships in close combat, but triereis needed also to be deployed, sometimes over long distances and in large numbers, into areas where conflict might take place, or at any rate where the threat of conflict might be a means of exerting political and economic power. Although the trieres was described by a contemporary writer as 'a rowing mill', there can be little doubt that voyages would have been made under sail whenever this alternative was as speedy, or desirable to rest the oarcrew. Oarsmen have always been easily exhausted by working for too long or in adverse conditions. A certain amount of evidence is available about the trieres' performance on a few particular voyages to distant theatres of war which can provide examples for comparison with the performance of a reconstructed trieres.

The most specific statement about the trieres' performance under oar is made by Xenophon. He says, (*Anabasis 6.4.2*), that the passage from Byzantium to Heraclea,' a city on the south coast of the Black Sea, (see *AT* p.104 Map 13)' is a long day's voyage for a trieres under oar'. It is perhaps worth noting that one manuscript reads ' a very long day '. The distance is 129 nautical miles which implies an average speed of 8 knots for sixteen hours at the oar, or, more likely, 7 knots for eighteen.

Thucydides records (3.49) two famous voyages, the second certainly wholly under oar, of Athenian triereis in 427 BC. The first, because of its melancholy mission carrying a death sentence for the male population of Mytilene, was taken at a leisurely pace. With a stop for the night and one for the midday meal on the second day, the voyage, of 184.5 nautical miles, probably took 48 hours. The second voyage, carrying the reprieve on the following day, was made under oar, in shifts, without the customary stops. The second ship arrived after, but not long after, the first. Since each started after a decision by the Assembly, say at midday, the time of the second ship at the oar would have been 26-27 hours and her average speed about seven knots or less. There were fortunately, Thucydides says, no contrary winds.

The voyages described have been of single triereis. Thucydides also tells (8.101) of a fleet movement by the Spartan commander Mindarus before the battle of Cynossema in 411 BC. He covered the 124 miles from Chios to Rhoeteum in the Hellespont in 18 hours at an average speed of just under seven knots. If the time at the oar had been shorter the speed of course would have been higher.

The speed of triereis can be estimated from the accounts of voyages, but not precisely since they depend on guesses about the length of a rowing day and the length and frequency of periods of rest. The accounts suggest that commanders in a hurry had to rely on the oar, but naturally they would have been grateful for a favourable wind which would allow them not only to rest their oarsmen but quite likely to make a faster passage. It would seem that seven knots was a good average for a day's rowing, and that,in fair weather 125-30 nautical miles could be covered under oar in a long day. Thucydides, in the Mytilene story, states, as something unusual, that the crew of the second ship rowed in shifts eating and sleeping in turn and keeping going the whole twenty four hours. There are two passages, one in Xenophon (*Hellenica 6.2.27-30*) and one in Polyaenus (*5.22.4*), a not very reliable writer of the 2nd century AD, speaking of the practice of rowing in shifts as not unusual. It is possible therefore that what Thucydides regarded as unusual was eating and sleeping on board and keeping up the rowing for twenty four hours rather than rowing in shifts. Nevertheless if rowing in shifts was a common practice on longer passages it is surprising that it is not mentioned more often. On the other hand, it may have been such a common practice that it was not worth mentioning.

If seven knots was a good average speed for a day's rowing, considerably higher speeds would certainly have been achievable in the short bursts required of the crews in battle situations, but the ancient sources provide no means of telling what they were. Experiment up to date with the reconstructed ship, with oars which are manifestly too heavy, has produced a speed of just 9 knots very briefly and it is estimated that in favourable conditions and with better oars the ship would be capable of about 10½ knots in short bursts.

1.5.THE ICONOGRAPHY OF TRIEREIS IN THE FIFTH AND FOURTH CENTURIES BC

1.5.1. Archaeology has produced information about the breadth and, more roughly, the length of triereis in the fifth and fourth centuries, and more recently determined the method by which their hulls were built. Epigraphy has yielded a great many facts about the manning, the oar system and the gear of triereis in the middle quarters of the 4th century, and in particular about oars (names of the three categories, lengths and numbers), masts and sails, and important items of gear such as *hypozomata* and *askomata*. Literature adds a certain amount of factual information of importance, but its principal contribution has been to provide a vivid picture of the trieres in action, both in battle and on voyage, showing the sort of ship the reconstructed trieres must manifestly be.

The final category of evidence is formed by the representations of oared warships from the seventh to the end of the fourth centuries BC in vasepainting, reliefs and coins. None of these are labelled, but in them a pattern of development may be recognised from early times to the period when the trieres was the type of warship from which navies were drawn.

1.5.2. The earliest representations of Greek oared galleys on vasepaintings (e.g.*AT Fig.22*) show the typical high upcurving stern and the prolongation of the keel into a forefoot. Some also show rows of hooked tholepins along the gunwale of what appears to be a narrow hull. Later vasepaintings (6th century BC) realistically depict single files of oarsmen in action, fifteen and twenty-five a side, answering to the names found in literature, pentecontors (fifty-oared ships) and triacontors (thirty-oared ships), e.g. *AT Fig.20*. An Attic bowl found in Thebes, dated 735-710, (*AT Fig.23*) seems to confirm a Boeotian attempt, recorded in Homer (*Iliad 2.509-510*), to more than double the oarsmen in a galley by sitting them at two levels; and at the very end of the eighth century a smaller two-level ship is shown in a relief depicting the evacuation of Tyre (*AT Fig.24*). In the early sixth century there are a number of black-figure

Figure 1 The Lenormant Relief

Figure 2 The Eleusis Relief

vasepaintings showing smaller galleys with oars at two levels, e.g.,*AT Figs.25 and 26*. The aim of the two-level design is clearly to obtain a better power/weight ratio than the single-level galleys afforded; and it appears that this was most successfully achieved by decreasing the length rather than by increasing the oarsmen in number. The additional level was placed in the former cargo space, i.e., in the *thalamos*, the hold beneath the *zyga*, the thwarts on which the Homeric oarsmen sat, and under which they stored their goods.

In history the final stage in the evolution of the oared warship as a fast, agile ramming weapon was marked by the trieres. Other still larger warships came later, but they were developed with a different tactical aim in view. The trieres evolved, to judge from the evidence already seen, from following, roughly, the earlier, and apparently unsuccessful, Boeotian pattern with 60 oars a side at two levels, adding a third level of a similar size through the agency of an outrigger. And this step appears to have produced a successful three-level ship, her success lying in the favourable power/weight ratio which made the added complication of the oar-system worthwhile. The trieres as shown by the naval inventories had on each side of the ship 27 thalamian oarsmen (in the hold),27 zygian oarsmen (sitting on the thwarts) and thirty-one oarsmen rowing their oars through the outrigger, called thranites. In any one vertical triad all the oars were of the same length, either 9½ or 9 cubits (4.2 or 4.0 metres, see p.5).

There are five representations of parts of oared galleys which derive from the heyday of the trieres as the warship on which naval power was based, the fifth and first half of the fourth century. These are accordingly likely to represent triereis and may collectively be recognised as such from the characteristics they exhibit :-

1. The Vienna Fragment of a red-figure vase of 450 BC. (See*AT Fig.43*)
2. The Lenormant Relief. Fragment(a) 401 BC (Fig.1),(b) Additional Fragment.(See *AT Fig.14*).
3. A lost relief, now represented by an early 17th century drawing in the British Museum, possibly a fragment of the relief to which 2. originally belonged.(See *AT.Fig.11*).
4. A relief fragment in the Museum at L'Aquila in Italy showing the stern of a galley similar to that shown in 2. and 3.(See *AT Fig.12*)
5. Detail of a painting on a red-figure vase of about 400 BC. in the Jatta collection at Ruvo in Italy.(See *AT Fig.41a.*)

In addition there are three further representations of later date:-

6. An Apulian rhyton in the shape of a ship's bow of 400-322 BC. In the Museum of the Petit Palais, Paris. (See *AT Fig.37.*)
7. A Coin of Cius in Bithynia 340 BC. (See *AT Fig.39*).
8. A relief in the Museum at Eleusis recently noticed by M.Lucien Basch. Date uncertain. (Fig.2 : see also *Le Musée imaginaire de la marine antique* (=MIMA): Athens 1987 p.290.)

All these items are works of art executed with varying degrees of realism. In 3. the reclining figure on the deck (the bow officer, *prorates*,) is larger than the oarsmen, and in 5 the size of the figures is greatly out of proportion to the ship. Attempts at perspective are unrealistic. In 1., 2., 3., 4., 5. and 7. the upper part of the the deck stanchions are shown as curved from side view, presumably to show a curvature which would not in fact be visible from the side . In 8. they are shown as straight. In 2. the lower end of the deck stanchions and the brackets for the outrigger are shown parallel to the oars, presumably to suit the artist's convenience, but in 5. where there are no oars, the upper part of the stanchions is shown curving in one direction and the lower part in the other, while the bracket for the outrigger is shown straight. In 8. the lower part of the stanchion and the bracket have become merged with the oars, which are of quite disproportionate length.

In spite of these sometimes confusing characteristics the representations in general confirm each other, showing a number of features which mark as triereis the ships partially depicted.

There is a **canopy deck** in all of them supported (except in 6.and 7.where the representation breaks off too soon) by stanchions resting (except in 1.,7. and 8.) on an external wale.

The **epotides** protecting the outrigger are clearly shown in 6. and the **outrigger** itself appears in higher relief in 2.,3.,4.,7. and 8. In 5., a vasepainting, it is shown in considerable detail and the ladder resting on it leaves no doubt that it projects from the ship's side. **Oars** are worked through it in 2.,3.,4. and 8. **Oarports** are shown at two levels between pairs of narrow wales in 1. and in detail in 5., where the lower ones are larger than the upper ones and apparently equipped with an **askoma** or leather sleeve. In 2. and 4., the Lenormant and l'Aquila reliefs, the thalamian oars emerge from a boss indicating the *askoma* and the zygian oars from under the outrigger, where an oarport may have been shown in colour in antiquity. In the case of 3., the dal Pozzo drawing, the 17th century artist has clearly misunderstood, and hence misrepresented, the oar system and the stanchions and brackets; but, the mistakes taken for what they are, the drawing is in agreement with 2. and 4.; 5., the Ruvo vase, exhibits the same details.

Number 8, the Eleusis relief, gives a much cruder picture of a trieres. There is a clear outrigger through which a file of thranite oars is rowed. The oarsmen are out of proportion to the oars in terms of the Lenormant and l'Aquila reliefs and their 'room' is unrealistically contracted. Two further files of oars emerge side by side from under the outrigger. It looks as if the sculptor has merged the outrigger bracket and/or the lower part of the deck stanchion with the thalamian oars. The literal interpretation of this relief leads to a mechanically impossible reconstruction (MIMA Fig.623).

The Vienna fragment, 1., is dated 50 years earlier than the Lenormant relief and the Ruvo vase. The curving deck stanchion, the pair of outrigger timbers and the two levels of oarports suggest that it represents part of the side of a trieres, probably near the bow on the starboard side. The differences it shows from 2.,3. and 4, are as follows:- the pairs of narrow wales on each side of the rows of oarports, the absence of a continuation of the deck stanchions downwards to the wales, the absence of brackets to support the outrigger, the presence of a substantial deck railing, the presence of vertical lines above and below the deck, and the decorative roundel of which part is visible. The first four of these are details which may be accounted for by the early date of the fragment which may thus portray an earlier design. The fifth may be explained by sails and rigging, and since no oars are shown the ship may be under sail. The fifth is paralleled by earlier ship pictures (eg.*GOS 12c*) and may be an indication that the part of the ship shown is near the bow.

CHAPTER ONE :THE EVIDENCE

There is a red-figure vasepainting dated between 500 and 480 BC in the British Museum (*AT Fig.47*) which depicts Odysseus and the Sirens. Although the ship shown is certainly not a trieres, she provides the only illustration of a sheath, presumably of metal, fitted over the forefoot of an oared vessel to form a ram. It also shows a truck, also presumably of metal, fitted at the masthead, through which halyards were worked. The oars too are of interest, since they appear to have fitted blades.

The preceding chapter has set out, in an order of scientific respectability, the evidence on which the reconstruction of an Athenian trieres of the fifth and fourth century BC has been built. In the absence of the remains of an undoubted trieres on the sea bed, all the evidence is a matter of argument, but in differing degrees of force. Nevertheless, when all the evidence is set out methodically, it is, I think, reasonable to claim that enough is known about the ship to have made the reconstruction a reasonable proposition. John Coates accepted it as such. To make a real ship on the basis of what was to a large extent a random, but not entirely inadequate, collection of information, was a quite remarkable achievement about which he will write in the following chapter.

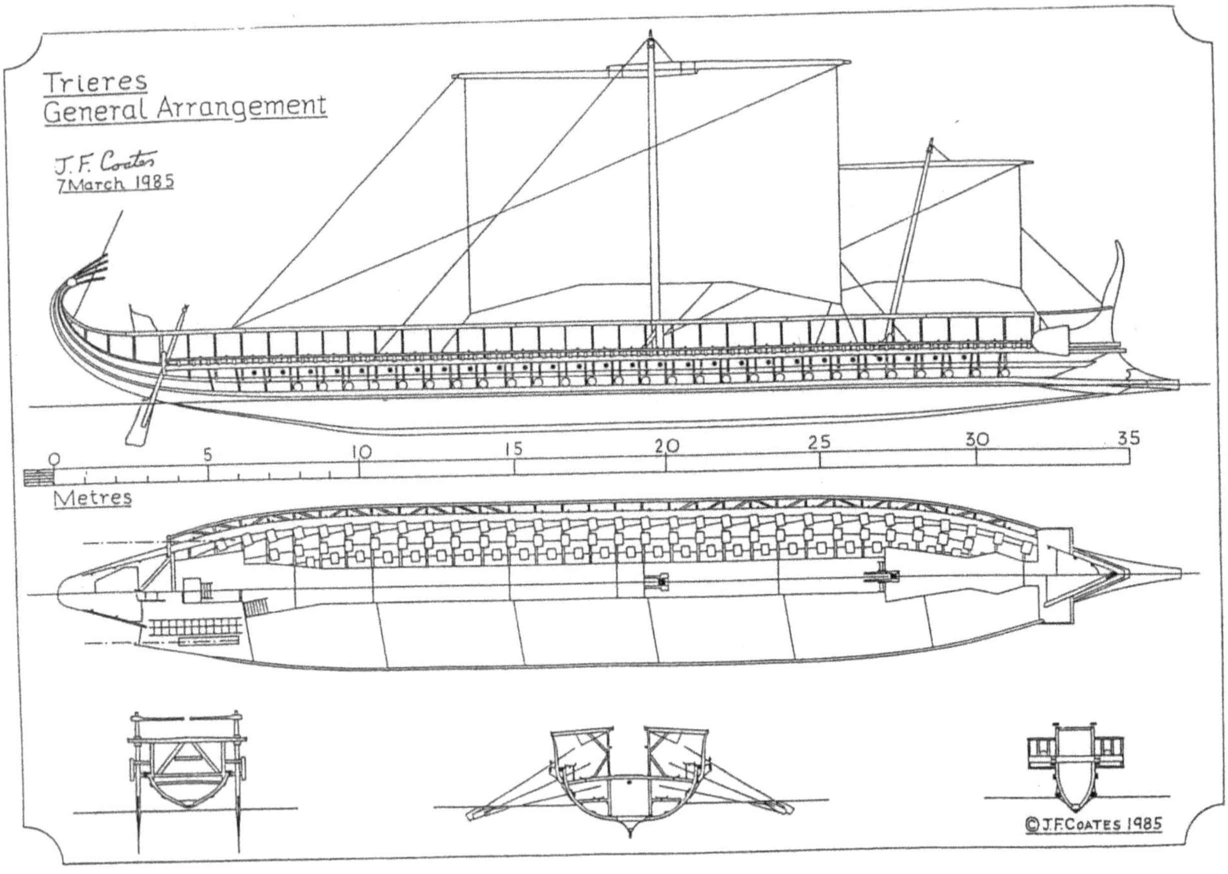

Figure 3 General Arrangement of Reconstruction

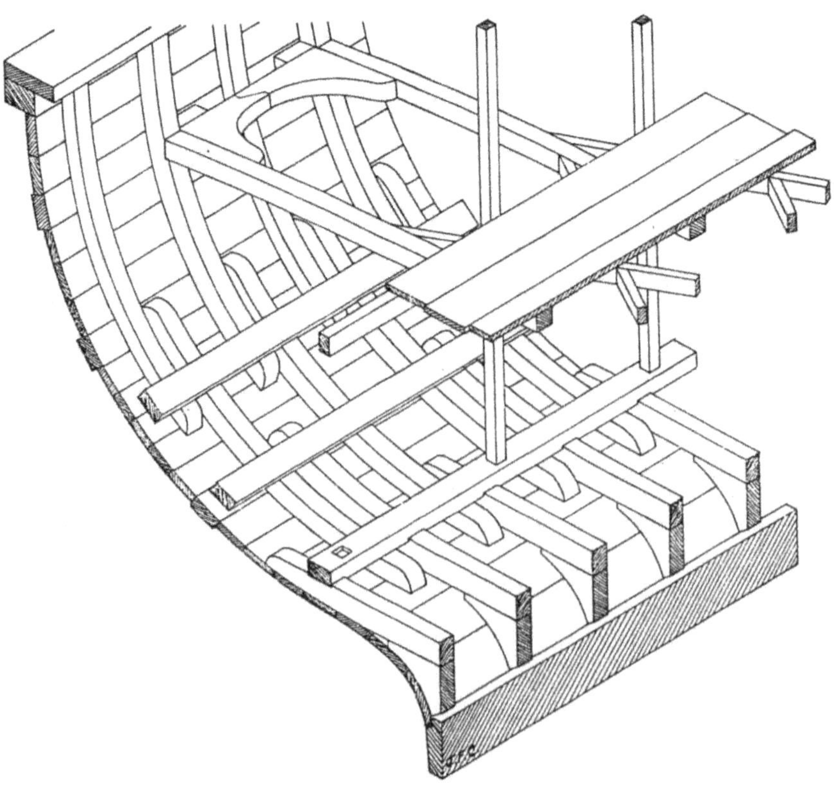

Figure 4 Hull Structure

CHAPTER TWO : THE RECONSTRUCTION

THE RECONSTRUCTION (1981-1987) : John Coates

2.1.DESIGN AND JUSTIFICATION

The evidence with which any authentic reconstruction of the trieres must accord has been rehearsed in the previous chapter. It need scarcely be said that to be authentic a reconstruction must also have a workable oar-system, float at the correct waterline to enable the oarsystem to work efficiently, be stable enough not only for working the oars but also to carry sail, and its hull must be strong enough to survive in waves of a severity in which the ship in other respects is capable of operating. Lastly it must be possible actually to build the reconstruction.

All these, which may be called physical requirements, assuredly applied as much in antiquity as they do today, since the laws of the physical world have not changed. Upon carrying out the design and calculation processes necessary to investigate the mechanics, weight, flotation, stability, propulsion and strength of the trieres, all were found to be closely interdependent. These essential aspects of the trieres also call for a compactness of oarcrew, weight, hull strength and stability, hard to satisfy simultaneously within the limitations of the materials and techniques available in antiquity. It emerged that the physical requirements when laid on top of the necessity to accord with evidence narrowed the scope of any fundamental variation between possible reconstructions to insignificance. It is upon that point that the authenticity of the reconstruction and the justification for building and operating it rest.

The point just made tends to apply particularly to artefacts of high performance near the limits of contemporary technical possibility, especially where elements of their design are interacting. In the ancient world, while such artefacts may not have been as common as in more recent times, the trieres may well have been an outstanding example. The point arises with less force in connection with merchant ships, as indeed in the bulk of artefacts studied by archaeologists today. Where it does apply it can be tested only by trying to arrive at alternative designs by drawing and calculation.

The nature of the interactions between elements of the design of the reconstruction may be indicated by quoting from *AT* pp.194-196:-

' Because the size of the trieres is for us virtually determined by the number and arrangement of oarsmen, and by the size of the ship-sheds (although of course in historical terms the reverse is the case), it was the arrangement of oarsmen which had to be settled first, sufficiently at least for their arrangement to be a basis for designing the ship as a whole. In doing this, design fundamentals immediately arose: attested speeds demanded an arrangement with a high oar-gearing of about three (see Glossary) as in most fast-pulling sea boats like gigs: likely power available required that the wetted area of the bottom be small, and that accordingly the beam on the waterline and the displacement be minimal. From that it followed that the height of the centre of gravity would be critical in achieving adequate stability without ballast. The high oar-gearing with oars of given length led to short looms, which did much towards enabling oarsmen to be closely arranged vertically and thus to a lowering of the ship's centre of gravity. The normal method of ship-building in the ancient Mediterranean, as demonstrated by recent underwater archaeological work, required a ship of the fourth century BC to have a wine-glass shaped mid-section. After much calculation and drawing, it turned out that a hull of overall beam fitting the ship-sheds and of that shape (*AT Fig.59*) with convex flaring sides and outriggers would :-

(a) house transversely without waste of space three files of oarsmen on each side tightly but workably packed by placing each man outboard of, and in height overlapping, the one below, provided that thalamian tholes were set inboard and their ports enlarged (as indeed they seem to have been in antiquity : see above p.7) to allow oar movement;

(b) house longitudinally the required number of oarsmen in each file in a fine enough hull of overall length just less than the dry length of the ship-shed slips, provided that the hull had a parallel middle body (*AT Fig.59*) and a large prismatic coefficient (see Glossary);

(c) provide sufficient displacement volume to float the total weight of the ship at the right waterline;

(d) have sufficient beam on the waterline to provide on that displacement adequate stability for sailing across a wind of about 15 knots, i.e., a metacentric height (see Glossary) of 0.5 to 0.7 metres.

"Thus all the evidence about the trieres could be satisfied by a design of hull and an arrangement of oarsmen in which physical considerations of hydrostatics, fluid mechanics, physiology and properties of materials, acting together, fix the beam on the waterline very closely, and the height of the centre of gravity, the weight and the hull form within fairly tight limits. It follows that any physically satisfactory reconstruction of a trieres conforming to the same evidence would have much the same values for these leading features. All are interlocked so that any fundamentally different solution is difficult to imagine.

"Speeds attested in antiquity seem only just attainable, so that any increase in weight or wetted area is out of the question. The centre of gravity could not be any lower without reducing the height of thalamian tholes above the water and thus rendering performance too sensitive to waves, heel or rolling. If the centre of gravity were higher the beam would have had to be increased to restore stability and that would have demanded either thalamian ports too large to fit between the necessary top timbers of the hull, or greater beam overall, too large for the sheds. If the prismatic coefficient were reduced to lessen wave resistance, without increasing the length, the foremost and aftermost thalamian oarsmen, despite short oars, could not have fitted inside the hull. If that coefficient were increased, wave resistance at high speeds would have risen to make it unlikely that the reconstructed ship could attain the attested speeds. Length could have been increased by a metre or two with possible gain at high speed but a penalty at more usual speeds, and the ram would then have overhung the water inconveniently when the ship was in the ship-shed."

In view of the critical nature of the trieres' design with her tightly-packed oarcrew, some experiments were needed to establish certain points about the oarsystem and to demonstrate that the hull could be built as specified. The experiments ranged from towing a model of the hull in the tank at the National Technical University at Athens, strength tests of the tenon joints between shell planks, models to verify the run of planking at the stern, a model of the ram structure and of the rig, a four-seater mock-up of the oarsystem, and finally to a trial section of the ship, built to the intended specification, with fifteen places for oarcrew and working oars. The trial section was put afloat and the oarsystem proved by practical test. These experiments, to which was added one to explore the behaviour of ropes when twisted into tension tourniquets, constituted a series of tests of critical elements of the reconstruction of generally

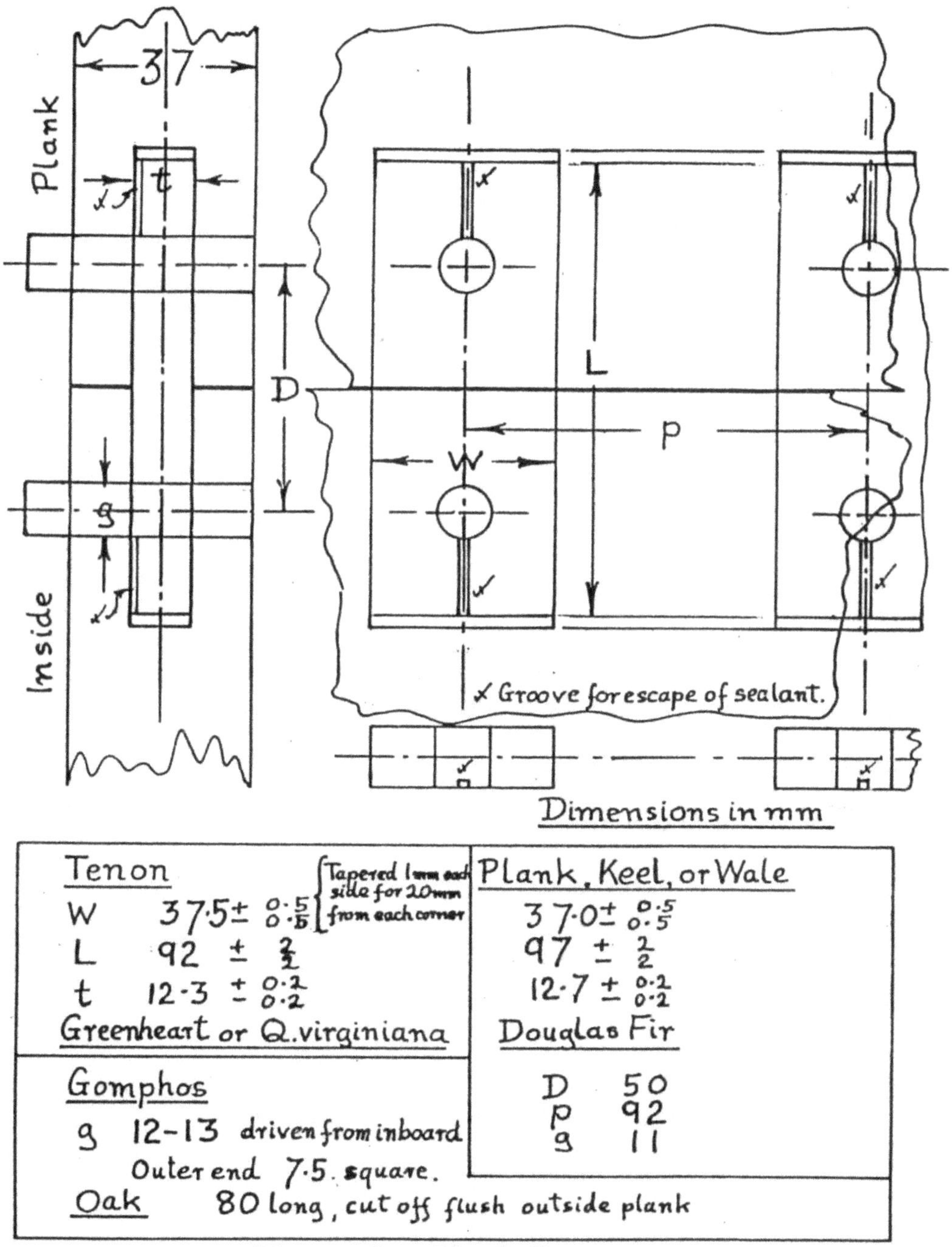

Figure 5 Plank Tenons

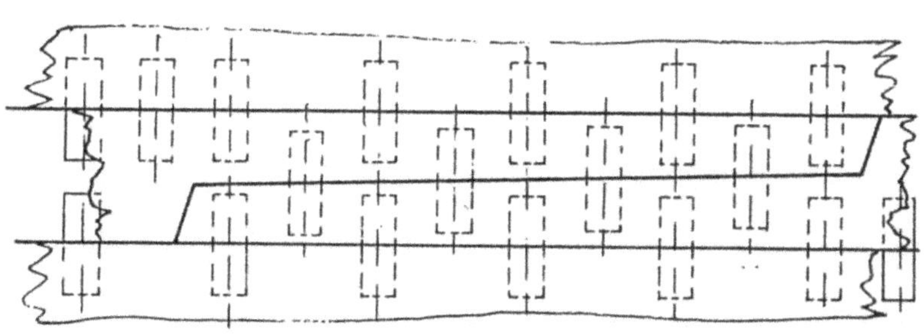

Figure 6 Plank Butt Scarf

increasing scope and cost as they proceeded, the trial section amounting to building a tenth part of the whole ship. The design and development programme thus ensured both the feasibility of the building of the ship and her subsequent performance sufficiently to reduce to an acceptable level the technical risk in committing the resources needed for the actual building.

2.2.MAIN PARTICULARS

The general arrangement of the reconstruction is shown in Fig.3. Without crew the ship weighs 25 tonnes. Fully manned the weight is about 40 tonnes. The main dimensions are :-

Length overall	36.8 metres
Breadth overall	5.45 metres
Length on waterline:40 tonnes	32.2 metres
Breadth " " " " "	3.62 metres
Draft at 40 tonnes	1.10 metres
Shell planking, thicknesses	37 and 55 mm
Framing timbers, spaced 444 mm	74 mm square in section
Keel	278 mm deep 175 mm wide
Beams, spaced 888 mm	92 mm square

Other particulars at 40 tonnes:-

Sinkage per tonne added	1.1 cm
Centre of gravity above underside of keel	2.2 metres
Metacentric ht.with yards hoisted	0.6 metres

The hull is built of iroko (*Chlorophora excelsa*) and Douglas fir (*Pseudotsuga menziesii*), instead of oak and a Mediterranean pine or fir, on account of the present-day lack of long straight timber in the Mediterranean area. The substitutes used have very similar properties to those of oak and pine. The keel, stem and ram timbers, as well as the framing are of iroko, the rest being of Douglas 'fir'. The oars are also of Douglas 'fir' instead of spruce or a fir, which has added significantly to their weight.

The structure of the ship is illustrated in Fig.4. The keel, planks and wales forming the shell of the hull are joined together by tenons (see Fig.5) closely spaced, as a rule 92 mm. centre to centre along the mid-thickness of planks. Every tenon, of which there are about 20,000 in the whole hull, is locked in place by an oak peg through each end, driven from inboard. The hull is of great length relative to its depth and consequently large forces are expected to be imposed upon tenons arising from the tendency of one plank to slide upon the next as the shell is subjected to bending and shearing actions when afloat and in waves. Accordingly the tenons are made of a particularly hard wood, *Quercus virginiana*, in preference to Turkey oak (*Q.cerris*) or olive (*olea sp.*)used in antiquity. This change was made to increase the operational life of the ship in consideration of her cost. As shown in Appendix 3, an interference fit was specified between tenons and their mortices in the longitudinal direction of the ship to obtain the greatest possible resistance to the sliding of one plank upon the next, and hence to the shearing of the hull in service.

The hull was built shell first, but in the absence of any present knowledge of the ancient method of creating and maintaining the form of the shell as it was built, and as that form was critical in the trieres to accommodate the oarsystem besides achieving the correct waterline and

Figure 7 Hull Interior

Figure 8

Beam Ends, Oarcrew Seats and Stretchers

adequate stability, the shell was formed round temporary moulds. The floors, futtocks and top timbers (see *AT Fig.64*) were fitted as the shell in way of them was completed. These timbers are fastened to the shell in the ancient manner, by 8 mm.square copper spikes, one for each plank-frame intersection, driven up the centre of Douglas fir dowels 18,5 mm. diameter, with the points of the spikes clenched over along and into the face of the timbers. Plank butts (see Glossary) are scarfed as shown in Fig.6, except in those places where they were used to make up longer planks, which were available to the ancients, but are not at the present time available commercially. Such longer planks are needed in the trieres particularly in the upper parts of the hull amidships where mechanical scarfs should be avoided on account of the high tensile stresses developed in that region of the hull owing to hogging (see Glossary) when afloat. They were made by joining planks together end-to-end by glued scarfs.

The long, undecked, hull would have been too weak in torsion without bracing between the beams (Fig.7) which is fitted throughout the ship wherever possible. For the same reason all beam ends are secured by lodging, i.e., horizontal, knees to the side of the hull (Fig.8): hanging, i.e. vertical, knees have been omitted. Beams, though forming a restriction for the thalamian oarsmen, are essential for the structure of the hull, besides being the *zyga* on which the zygian oarsmen sit.

2.2.3. *The Oarsystem*

This is shown in Figs 8, 9 and 10, the last indicating the motion of the oarcrew through the stroke of the oars and the maximum obtainable length of the stroke of the handles, which is not far short of the 888 mm.spacing of the oarsmen themselves. The argument by which the oarsystem was derived from the evidence given in Chapter 1 is presented in more detail in *AT pp.212-218*. As this reconstruction was designed and built to a 'room' (Gr.*dipechiake*, Lat. *interscalmium*) of exactly two Athenian cubits of 444 mm., leg-room is limited and this has consequences for the selection of oarcrew from present-day populations.

2.2.4. *Hypozomata*

In the reconstruction these tensioned ropes are rigged between two anchorages on the middle line of the ship. They run parallel and together uninterruptedly just under the beams (Fig.8) about 0.6 m. above the flexural neutral axis of the hull(see Glossary). The effect of the tension upon the hull is to impose a longitudinal compressive force at a height such that the stress so generated in the gunwales is about 1 Newton/mm^2 compressive diminishing linearly with depth from that level down to nearly zero at the keel. The *hypozomata* thus reduce tensile bending stresses in the gunwale caused by hogging by about 1 N/mm^2. These ropes do virtually nothing however to alleviate shear forces in the shell unavoidably associated with hull bending.

It was intended that the *hypozomata* should be made of two loops of rope 40 mm. in diameter,each loop passing round and between the anchorages at each end of the ship, the whole then being twisted together by a transverse bar amidships passing between the two pairs of ropes so formed, to make a tension tourniquet nearly as long as the ship. The behaviour of tension tourniquets of rope had to be investigated, as reported in IJNA 1987 16.3 pp.207-211. In the event,shortly before launch, difficulties arose with the mechanism controlling the twisting bar. To get the ship operational in the short time left before the trials which are the subject of this report, it was decided temporarily to abandon the twisted ropes and to achieve the same effect by the use of a wire hawser tensioned to 12 tonnes.

...

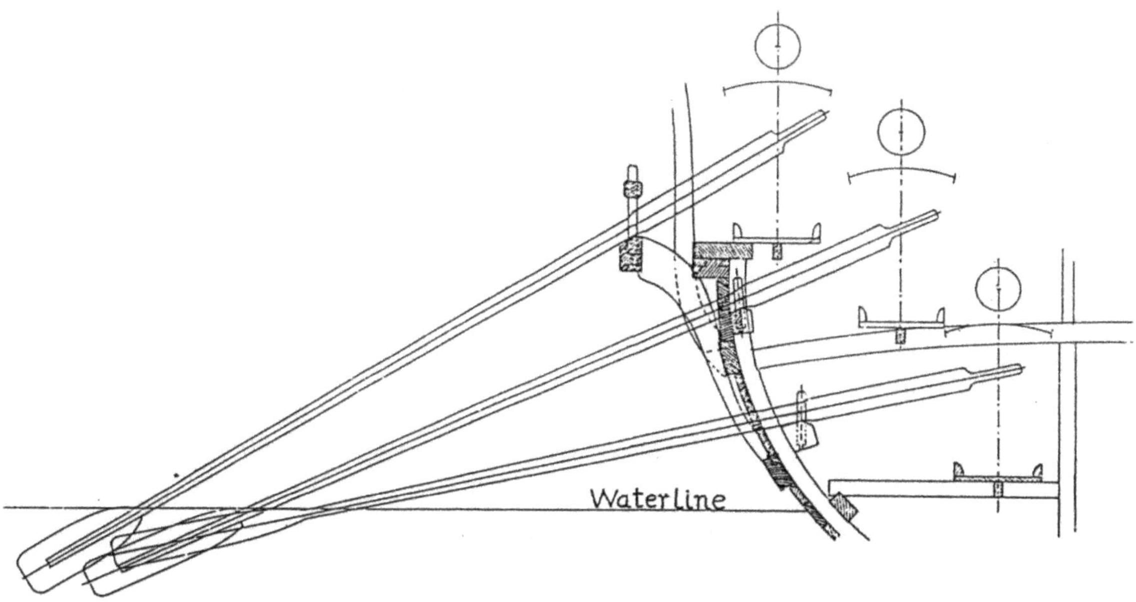

Figure 9 Oarsystem

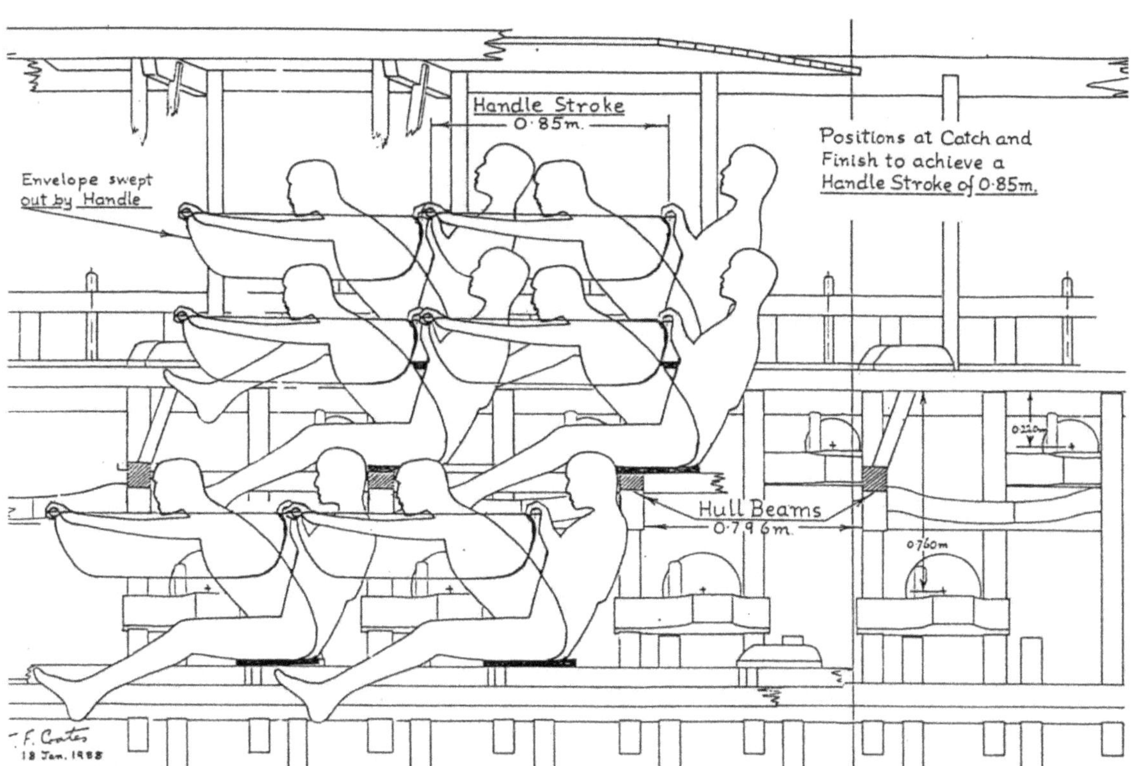

Figure 10 Oarstroke Diagram

2.2.5.*Rudders*

The two rudders (see *AT Fig.62*), each 5 metres long with their stocks of diameter 180 mm., are made of Douglas fir. They are carried in bearings at the after ends of the outriggers and where the stocks pass through the Quarter Deck. The height of each rudder can be adjusted by halyard, and they can also be hinged about their lower bearings to raise them to reduce their resistance to the water or for mooring the stern to a quay or beach.

2.2.6.*Rig*

The sail and rigging plan is given in *AT Fig.68*, where the brailing lines and braces of the foresail have been omitted for clarity. The mainsail has an area of 95 sq.metres and the foresail has the same proportions but half the dimensions. The mainmast is 11.2 m. from heel to truck and the main yard, made of two spars fished together, is 13 m.long. The mainmast is stepped in a tabernacle about half a metre forward of the mid-length of the waterline: it hinges at beam level for raising or lowering.

The sail area was designed to cause the ship to heel about 7 degrees in a steady beam wind of 15 knots, under mainsail only. The zygian oarports which are open become submerged, so that the ship would be swamped, at a heel angle of 21 degrees.

The rig was designed by Owain Roberts and is more fully described in the *Proceedings of the Second Symposium on Ship Construction in Antiquity, Delphi 1987* (to be published by the Hellenic Institute for the Preservation of Nautical Tradition, Piraeus).

2.2.7.*Features not resolved and departures from authenticity*

2.2.7.1 Unresolved features include :-

(i) The fit of hull shell planking tenons which were not fitted tightly in the longitudinal direction, allowing some hogging by shear to occur.

(ii) The material and design of the oars which need refinement on the basis of data to be obtained in further trials.

(iii) Achievement of a plausibly authentic method of rigging and tightening the *hypozomata*.

2.2.7.2 The principal departures from authenticity, some of which have been already mentioned, include:-

(iv) The use of Douglas fir and iroko in building the ship.

(v) The use of modern glues to join some planks and certain longitudinal timbers end-to-end to make members as long as those recorded in antiquity in connection with building warships.

(vi) The use of temporary moulds to define the shape of the hull, and of power tools in building the ship.

(vii) The use of sawn instead of grown timbers for the hull framing.

2.2.7.3 Other departures are:-

(viii) The sealing of planking seams underwater with a modern two-part sealant, and the protection of the underwater body by modern paints. The protection of timbers generally by modern timber preservatives.

(ix) The use of modern fibres in ropes.

(x) The use of iron anchors of 19th century design.

(xi) The use of circular, instead of square or lozenge-shaped clench, rings under heads and points of hull bolts.

(xii) The use of modern navigation instruments and the fitting of navigation lights to comply with present-day regulations.

(xiii) The carrying of inflatable buoyancy bags for crew safety in the event of swamping.

(xiv) The gluing together of the two spars forming the main yard, and the temporary fitting of mast shrouds.

Departures from authenticity have been discussed in greater detail in *Antiquity* Vol.61 No.131 March 1987.

The performance of the reconstructed ship is not thought to have been affected by any of the above except (ii), the material and design of the oars.

CHAPTER THREE : THE SEA TRIALS

THE SEA TRIALS OF OLYMPIAS AT POROS : AUGUST 1987

IAN WHITEHEAD, JOHN COATES, OWAIN ROBERTS

3.1.OBJECTIVES

3.1.1.To familiarise the oarcrew with the ship and sea oars.
3.1.2.To learn essential procedures for starting and stopping when pulling and backing, and for turning under oars.
3.1.3.To learn to pull the trieres at various striking rates.
3.1.4.To measure the distances which can be travelled in various periods of time.
3.1.5.To measure acceleration from rest, ahead and astern.
3.1.6.To measure the time taken to turn 90° and 180° under oar, from rest and from various speeds ahead.
3.1.7.To measure rates of turning by rudders at various speeds.
3.1.8.To measure the performance of the trieres as a sailing ship.

3.2.THE SEA TRIALS

3.2.1 *The ship*

The trials showed that the reconstructed ship was workable as regards flotation, stability, propulsion and manoeuvrability under oar, and, so far as it was tried, under sail. As only slight sea and swell were encountered, the strength of the hull in longitudinal bending was not tested. The oars were shown to be too heavy and unbalanced to allow more than half the possible power, at most, to be developed. Manoeuvrability, taking account of the length of the ship, was very good.

3.2.1.1. *Weight, trim and stability* The displacement of the ship, fully manned, was at first 40 tonnes. During the second half of the period displacement was increased to 45 tonnes by ballasting to adjust the waterline to its design level suitable for better oar-handling (Annex 1).She was designed to carry some additional equipment which was missing, e.g. water and other stores, buoyancy bags, lifejackets, sleeping bags, navigation equipment etc., and as a result the bow oarcrew were about 10cm too high out of the water, before the ballast was shipped.

<u>Drafts during trials in metres</u>

	Before ballasting				After ballasting			
	Unmanned		Manned		Unmanned		Manned	
	P	S	P	S	P	S	P	S
	mean		mean		mean		mean	
F'd mks	0.87	0.84	1.05	1.02	0.95	0.93		
	0.855		1,035		0.94			
11 Ord.	0.91		1.093		0.973			
Aft mks	1.01	0.97	1.19	1.17	1.04	1.025		
	0.99		1.18		1.023			
Displacement	24.3 tonnes		39.9 tonnes		29.4 tonnes		45.0 tonnes	

<u>Longitudinal positions of draft marks</u>

Forward marks: 6.46m forward of 11 ordinate (mid-length of waterline at 49 tonnes)
Aft marks : 9.69m. aft of 11 ordinate

Weights

Weight of crew and observers 15.6 tonnes
Weight of ballast 5.0 tonnes spread between top timbers 23 and 33,viz.25 bags of stones at 20 kg.each

L.C.G.(see Glossary) 2,28 m. forward of 11 ordinate. The foregoing drafts, after ballasting, refer to the foregoing position of ballast. Twenty bags were subsequently moved ten Top Timber spaces (i.e. 4.44 m.) aft to bring the lower wale, and hence the files of oarcrew, parallel to the waterline.

Stability proved to be ample and the steadiness of the ship led many observers and oarcrew to suggest that the beam on the waterline was unnecessarily great (3.70m at 45 tonnes displacement). When inclined, at 37.5 tonnes, G.M.(see Glossary) was found to be about 0.8m. (Accurate figures for the inclining operation are held by the Hellenic Navy Dockyard at Salamis). In the conditions of the trials the G.M. of the ship with yards raised is estimated to have been about 0.62 m.. Annex 1 should be consulted for the hydrostatic curves of the ship and an approximate curve of G.Z.(see Glossary) under trial conditions with a displacement of 45 tonnes, also for a statement of the principal weight groups.

In spite of the perceived steadiness of the ship, it was found to be necessary to dispose passengers on the canopy-deck more or less evenly to port and starboard. Heel was thus kept to within 1°-2° to avoid disturbing the work of the oarcrew. Movement across the deck had similarly to be controlled. This experience underlines the greater stability and therefore waterline beam needed in warships built to carry troops (on deck) compared with fast triereis such as the type reconstructed.

3.2.1.2.*Structure* After about a month afloat the hull had hogged (i.e.the ends had dropped relative to the middle) by 55 mm.. Over the period of the trials the hog did not increase, but the hull was not subjected to significant bending by sea swell during that time. The matter is discussed in Annex 2.

Leakage was only slight, the depth of bilge water above the keel being usually kept to about 15 cm. at the after cut-up of the keel while the forward cut-up was dry. The volume of that amount of water is about 0.2 cubic metres. The small leakage has no historical significance because the seams of the reconstructed ship had been sealed with a modern two-part polysulphide sealant.

Apart from the suspected looseness of the plank tenons (Figs.5 and 6) fastenings of the main hull structure appeared to be satisfactory. A few built-up floors right forward had broken at joints, but there seemed to be no pressing need for their repair.

The torsional stiffness of the hull, which, since the hull is undecked and long, had given cause for some concern, seemed adequate at least for the activities and conditions of the trials. No working of the beam lodging knees or beam braces (Fig.4) was apparent.

3.2.1.3. The *hypozoma* (Fig.7) Tension of the *hypozoma* during the trials was a little under 12 tonnes. Though a higher tension might have proved better for the hull by reducing hogging, it was felt advisable not to raise it because the forward anchorage (in the form of a built-up breast-hook lodging between and checked into the inside surfaces of the middle wales, just aft of the stem) was showing signs of spreading transversely, possibly owing to a suspected absence of certain through-bolts. The after anchorage, already modified, seemed to be satisfactory.

3.2.1.4. *Oarcrew's furniture* This was a source of trouble during the trials, mainly from drawn nails and loosened tholepins. The stretching and breaking of the leather oarloops which were both too thin and had been weakened by badly arranged stitching are the subject of discussion elsewhere in this report (Annex 3). Oarcrew foot-stretchers (Fig.8) proved satisfactory. Their design had been strengthened on experience gained in the Trial Piece at Henley in 1985, when about a third broke or split in the course of use.

3.2.1.5. *Oar System* The oars proved too heavy. Combined with the crew's lack of training,fitness and uniformity of body-size,the weight and lack of balance of the oars limited propulsive power to less than half that which a crew of 170 should be able to deliver. The oars (Fig.9) are of Douglas fir, heavier than the timbers specified but chosen for its better availability in Greece. The diameter of shafts and thickness of blades were chosen to reduce risk of breakage during early operation of the ship(when everyone concerned would be feeling their way) without undue loss of performance.In the event six oars were broken mostly in circumstances in which they would have been broken no matter how strong they had been (within practical limits). Conversely few, if any, more would have been broken had they been lighter and less robust. Oars are discussed further elsewhere in this chapter. The Hellenic Navy has undertaken to counterbalance or lighten the oars, as well as to mitigate the effects of ill-fitting plank tenons, as suggested in Annex 2, together with other minor modifications required in the ship before further trials are undertaken.

3.2.1.6. *Sails and Rig* These were not tested thoroughly during these trials. The basis of the design is described by Owain Roberts in the paper referred to above (p.24). Although the trieres was a sailing galley, for which performance under sail would have been important in making passages over long distances, it was agreed to give sailing a low priority in the first trials, so that the historically important and still largely hypothetical oar system could be given an extended trial.

The main mast and 'boat' mast (see Glossary) had been given two pairs of shrouds each contrary to the rigging plan supplied, and fitted in a way which interfered with the working of the sails and to some extent with the rowing. There is no evidence for such shrouds in antiquity. It is understood that they will be removed in time for the 1988 trials. Certain other anachronistic modifications of the rigging plan had also been made.

The main yard also had not been built to specification. Instead of two spars of solid wood 'fished' together and thus admitting some flexibility between the two halves, the yard consisted of two halves built up of laminations and finally glued together. In cross section all dimensions had increased by 25%, so that the yard was grossly overweight and inflexible. It required a determined crew to hoist it. A further departure from specification was the use of large brass screws to fasten the brail fairleads to the yard.

The sails Flax canvas obtained in Britain had been made up with little knowledge of sailmaking. Departures from the sail plan included the use of brass punched-eyelets, insufficient sewing and a poor quality of leather for the tabling which ripped when lightly stressed. The result was that on two separate occasions one clew and one tack pulled out, both with part of their reinforcing patches. Neither was more than lightly stressed.

Figure 11

Rowing Master's Position

Figure 12 Part-view of Oarcrew

Belaying pins and cleats These had not been fitted in the shipyard, but the Captain arranged for them to be made and fitted at Poros. The cleats for the halyards were fastened strongly to deck beams in the gangway, but the site chosen for the brail belaying pins was not so happy. They were placed on two separate rails at some distance from each other, and only lightly fastened at canopy-deck level to some of the supporting stanchions. They should have been brought together aft and most securely fastened to the hull structure.

3.2.2. *Command and communication*

In the classical period at Athens the command structure aboard triereis appears to have been simple. The trierarch was in overall command, while the *kubernetes* (helmsman) was in charge of navigation and when the ship was under oar gave orders to the *keleustes* (boatswain) whose job it was to manage the oarsmen and to get the best out of them (Plato *Alcibiades I 125 C :AT p.112*). Aboard the *Olympias* the command structure was nothing like as simple. For the carrying out of the trials programme the agreed procedure was that command should pass from the Captain to the Rowing Master once *Olympias* had left her mooring, and command remained with him unless the safety of the ship was threatened. On her return to her mooring the Captain took command for the final berthing manoeuvres. Since the Helmsman thus came under the command of the Rowing Master when the ship was at sea, this was a radical departure from ancient practice. The departure was perhaps justified by the circumstances of the trials, but had disadvantages. From time to time the Rowing Master had to come up from the gangway below to see how much room there was for a given manoeuvre, and sometimes it was difficult for the Helmsman or Captain to attract the attention of the Rowing Master because the latter was in the stern and faced forwards when giving commands (Fig.11). In antiquity the *keleustes* (see *AT* Fig.31) stood in the middle of the ship facing aft so that he was in direct visual contact with the Helmsman.

To complicate matters further there was, in the trials, the matter of language. Since the Captain spoke only a few words of English, and the Rowing Master had even fewer words of Greek, an interpreter was needed to make communication possible. A member of the Greek deck crew and a Greek member of the oarcrew did an excellent job between them, but the problem inevitably caused delay, and made it more difficult for the Captain to exercise command. The interpreter from the oarcrew, Yannis Manuelides, who had rowed in Britain and in U.S.A., was particularly useful since he could translate Greek not only into English but into boat-club English. The Helmsman, Timothy Shaw, also spoke no Greek. Eventually, the Captain gave the Helmsman control of the mooring manoeuvres which enabled the latter to carry them out effectively.

The words of command to the oarsmen in the trials were those of British river rowing and were of limited value in an oared warship. Terms such as bow side and stroke side were meaningless and apt to confuse. Port and starboard would have been more suitable but were unfamiliar to most of the oarsmen. The Hellenic Navy uses *aristera* and *dexia*, left and right.

The communication of commands from the Rowing Master to the oarsmen was another problem which was never satisfactorily resolved. The system adopted was barely adequate, and many complaints were registered in the questionnaires which were issued at the conclusion of the trials. The Rowing Master stood, or crouched (Fig.11), at the bottom of the steps

Figure 13 Angle of Thranite Oars (a)

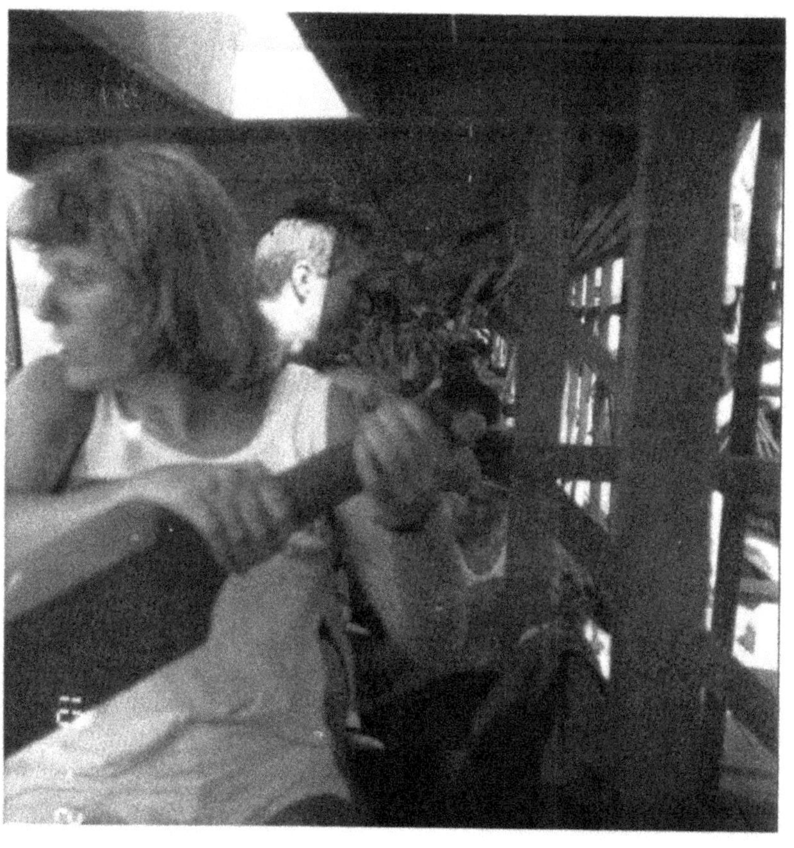

Figure 14 Angle of Thranite Oars (b)

leading to the gangway and directly forward of the Helmsman, issuing commands through a loudhailer. A deputy crouched by the mast and relayed commands forward by loudhailer. There was a time delay of about one stroke from stern to bow. The thalamians in the bow, placed below the platform occupied by the bow section of the deck crew, found it particularly difficult to hear commands, especially when that section was engaged in the mooring operation. If strict silence could have been kept there would have been fewer problems, but the system itself was inadequate, at any rate with a novice crew in a strange ship. Any sort of noise interfered with the communication of commands. Early in the trials oarsmen complained that the noise made by people walking about on the canopy-deck prevented them from hearing commands. Talking among the crew was another source of interference. Various attempts were made to sing in time with the stroke, but no song tried fitted the rhythm, and some found that the different timimg disrupted their stroke. During the song oarsmen up in the bow were even more isolated than usual, and were unable to hear any commands at all.

There were however some communications successes. The whistle signal used to call for an emergency 'easy oar' when a thalamian had been trapped after 'catching a crab' could be distinctly heard all over the ship. The high-pitched 'bleep' of the loudhailer also penetrated well. This experience suggested that the *auletes*, piper, a regular member of the non-rowing crew, (*hyperesia*) of an ancient trieres, may have been an important assistant of the *keleustes* in the transmission of commands to the oarsmen and not just a means of keeping up the striking rate in a long stint at the oar. The Rowing Master used the rhythmic chant *rup-pup-pai* quite often and this fitted the stroke well. Since Aristophanes identifies oarsmen with their chant as the *rup-pup-pai* in the *Wasps* (909), the chant is historically attested as well as being useful.

3.2.3. *Propulsion*

The intended programme of the trials was modified in the light of circumstance. The programme actually carried out is given in Annex 4.2.

3.2.3.1. *Rowers and rowing* Ancient sources make it clear that if a trireme was to be deployed successfully it was necessary to muster a crew of skilled and fit oarsmen. In Thucydides Brasidas is shown rallying his crews by playing down the importance of the greater experience and skill of the Athenians, and in Aristophanes a chorus of naval veterans declare that in their day all they were interested in was who could row the best (Thucydides 2.87, Aristophanes *Wasps* 1096). Oarsmen had to be trained if they were to reach the required standard. Athenian crews were sent out on active service for as much as eight months to learn and practise seamanship (Plutarch *Pericles* 11). The problem facing the Rowing Master at Poros in 1987 was that, without any experience of trieres rowing himself, he had to train a crew, also without any experience of the trieres, in the space of two weeks. Moreover during that two-week period a trials programme had to be carried out to assess the performance of the ship. How the Rowing Master approached these problems, how he and his crew fared in overcoming them, and what was discovered about rowing *Olympias* are the subjects of this part of the Report written by the head of the recording team, supplemented by Annex 8, which is of a descriptive nature and written by a member of the oarcrew.

Figure 15 Thranite Recovery

Figure 16 Part-view of Oarcrew at Mid-stroke

CHAPTER THREE : THE SEA TRIALS

The recording team was mostly preoccupied with measuring the performance of the ship during outings. Although several of us rowed during outings for short periods towards the end of the trials, this limited experience was certainly not enough to enable us to report authoritatively on the rowing of the ship. Originally it had been planned to issue questionnaires to some rowers after each outing,but this plan was abandoned. It was felt that such frequent sampling would prove tiresome to those chosen and would produce unsatisfactory results. Consequently, although we did canvass the opinion of ten rowers after outing 9, it was left until the end of the trials to issue questionnaires to the entire crew. Sixty completed questionnaires were returned, approximately 30% of the crew filling them in. This provided a reasonable sample of the rowers views, which could be studied at leisure and used to draw a balanced picture of how the rowing had gone. A specimen questionnaire is given in Annex 6. Dr Budd compiled a report and Graeme Fife, who rowed a thranite oar from the beginning of the trials right up to and including the Commissioning Ceremony in Piraeus on August 26th, provided a detailed account of his experiences. Both have been incorporated in this part of the Report.

3.2.3.2. *The crew of rowers* The rowers were predominantly students although there was a sizeable minority from non-college rowing clubs around the country. In the ship they were divided into six teams of about 35, with an equal weighting of experience within each team: 4-5 veterans, 8 or 9 women, and the remainder young oarsmen. All the rowers were volunteers paying their own expenses except for the accommodation which was provided free by the generosity of the Hellenic Navy. A teamleader/coach was appointed for each team. The bow teams, White and Yellow, rowed 27 oars apiece, and so did the stern teams, Red and Green. The amidships teams, Blue and Black, rowed thirty oars each. In theory there were at least five spare rowers for each team at each outing. In practice this was only true on two or three days, since a number of the rowers seemed to lack the necessary strength, fitness and to a lesser extent the skill, to handle the heavy oars. Unfortunately it was necessary to drop them because of the danger both to themselves and to their colleagues rowing near them. *Olympias* proved to be a tough ship and a certain basic level of fitness and strength was needed to row her. In Athens in the fourth century a trierarch claimed to have sent home the conscript oarsmen the city sent him. By contrast, other crew members complained that the outings were not long or fast enough.

As far as possible teams were selected in such a way as to group friends together. Rowers were asked to express a preference for bow ot stroke side, and were placed on the side they had chosen. Teams occupied the same part of the ship for all the outings except one, on which the bow and stern teams were switched. Discipline on board was mostly good. For reasons of safety order had to be maintained, and the occasional outbreaks of unruly (or'spirited') behaviour annoyed both the Captain and other rowers.

Paul Lipke adds:' Although many rowers could and did move around, rowing different levels and parts of the ship on different outings, most developed a preference for a particular position based on body-size, camaraderie with other oarsmen in their triad (see Glossary) and the idiosyncrasies of a given position.

Health Oarcrew health remained good throughout the period of the trials. Despite the great heat (temperatures : a.m.80 F,p.m.90 F) there was only one mild case of heat exhaustion. Rowers either drank water, or a mix

Figure 17 Three Triads of Oars from Above

Figure 18

Portside Oars at the Catch

of one sachet of Rehydrate/Dioralyte in 1½ litres of water, taken aboard regularly in individual bottles. Fluid consumption varied greatly. One rower took no water at all during a two-hour outing while others would consume the entire 1½ litres. Eight or nine people developed diarrhoea and dealt with it using their own remedies brought from home. Only one required medical attention.

There were several cases of heat or sweat rash, some made worse by exposure to the sun. Blisters were not much more of a problem than normal among rowers. One or two complained of chafed bottoms. Longer outings would probably have resulted in more cases of the latter two complaints. One rower remarked that Dr Budd's duties as both Rowing Master and Medical Officer were in conflict, since as the former it was not in his interest to pay attention to ill-health and distress among his crew.'

Rowing The first few outings were notable for the sound of clashing oarblades as the rowers struggled to master the heavy oars and unfamiliar technique of rowing at sea on fixed seats (Figs 8 and 12). Everyone, rowers and coaches, were in the dark as to how best to go about learning to pull *Olympias*. By the end of outing 1 it had become clear that the biggest and strongest rowers should be placed on the thranite (uppermost) level since the steep angle of the oar handle (Figs.13 and 14) and the weight of the oars made the recovery awkward (Fig.15). At about the same time it was realised that if a thalamian (lowest) rower caught a crab the consequences could be serious. The unfortunate rower's neck could be pinned by the oar handle against the transverse beam which ran behind his head, and the momentum of the ship would make it necessary to exert considerable force to release the victim. The thalamian position was the most cramped of the three and large rowers found that they had insufficient room to pull a full stroke. Consequently, small but skilled rowers were moved to that position, and the biggest and strongest were concentrated on the thranite level.

By outing 4 one of the major reasons for the thalamian oarsmen 'catching a crab' was seen to be that their blades were being knocked down by the blades of the zygian (middle) oarsmen. Although the zygian could get his oar out at the finish,no doubt assisted by the efforts of the thalamian below, the thalamian was unable to avoid, in this situation,'catching a crab'. Both the thalamian and the zygian oarsmen rowed without being able to see what was happening. The length of the thalamian stroke was limited by the distance the heads of the rowers in that position could travel between the beams of the ship, but in fact that distance was fully used by comparatively few thalamians,who thus shortened the stroke that could be achieved by the crew as a whole. It was further seen that the zygians needed to match their stroke length to that of the thalamians, both sets of oarsmen coached by the thranite oarsmen above them who alone could see what was going on. The thranites in their turn needed to look for the slot (Figs.16 and 17) between the zygian and thalamian blades and place their blade in that slot in the water. This practice would make the thranite responsible to some extent for the safety of the thalamian below and aft of him. By ensuring that his blade separated those of the zygian forward of him and the thalamian aft of him he compensated for their mistakes of timing and variations of stroke length, and thus reduced the number of 'crabs' caught. From the moment of this realisation progress was dramatic. Power increased (Fig.18) with very little blade-clashing and fewer incidents of crab-catching. It was still possible for the thalamian blade to become entangled with the zygian's aft of it, in the <u>next</u> triad, but the problem was now less serious. As can be seen from Fig.19 the triads did not match those indicated in the Lenormant relief (Fig.1 and *AT Fig.72*).

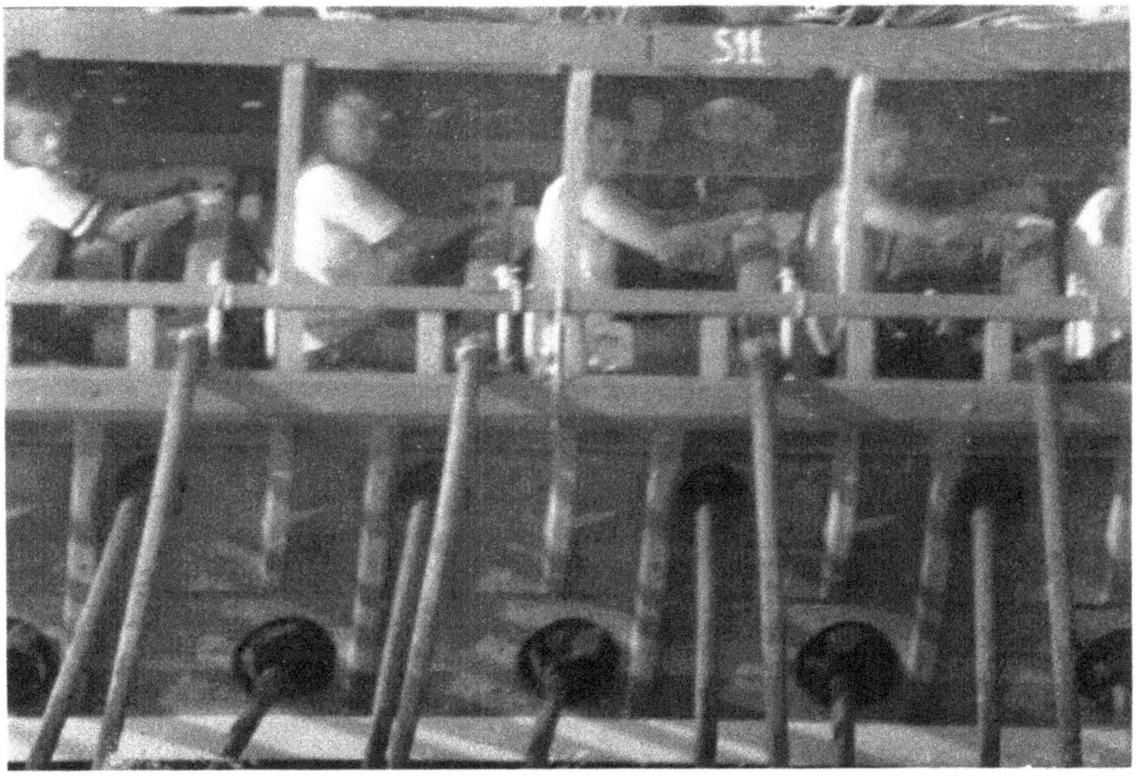

Figure 19 Close-up of Portside

Figure 20 Zygian Oarloom and Thranite Stretcher Rail

CHAPTER THREE : THE SEA TRIALS

As rowers became stronger and more skilled, mechanical failure (see below Annex 3) became the biggest obstacle to steady progress. It has to be said, however, that the length of handle stroke generally fell far short of the 0.8 m. permitted by the layout of the ship (Fig. 8) and necessary to achieve full power.

On outing 19 the two teams in the bow changed places with the two teams in the stern. The bow oarsmen now rowing in the stern complained that they now had less room for rowing, in particular the thalamians. They also remarked that since they were rowing in water disturbed by the oars forward of them they were less able to apply their work. The stern oarsmen now rowing in the bow found it much harder work. The greater room in the bow and the undisturbed water in which they were able to row made the bow section the most efficient in the ship. This conclusion was confirmed by the speed measurements made when the crew was rowing a section at a time. The five tonnes of ballast loaded forward of amidships after outing 11 corrected the rowing height in the bows. Previously some zygian blades would hardly go into the water, their looms being prevented from rising high enough by the stretcher rails of the thranites (Fig.20).

The responses of crew members to the questionnaire show clearly the great difference between rowing *Olympias* and sport rowing with which they were familiar. Almost everyone wished for lighter, better balanced, oars, and such oars would certainly have been used in antiquity. A downward force on the oar handle of between 10 and 13 kgs was required to recover the blade from the water. Lanyards soon appeared, to secure handles with the blades clear of the water and give the rowers a rest when they were not rowing. A few rowers had their wish for better balanced oars met by nailing strips of lead, as an experiment to the counterweight blocks of their oars. This remedy improved their balance and made recovery less tiring, but there is no trace of such practice in the ancient evidence, and it is no proper alternative to the provision of lighter oars.

Oarloops which stretched also caused problems and reduced the efficiency of the stroke. In this connection it is interesting to remember that each oarsman in antiquity was responsible for his own oarloop (*AT p.136*).

Some oars tended to twist off the square and the effort required to keep them at the correct angle made rowing more difficult.

The sheepskin cushions used as padding for the seats were welcomed by most of the crew, though some were happy to sit on the bare wood. The cushions had to be carefully dried out after each outing if they were not themselves likely to become a health hazard.

Only one or two of the most mature crewmembers had had experience of fixed seat rowing. The body swing required (Figs.9, 21 and 22) to give a stroke of decent length when rowing in a fixed seat vessel was alien to the great majority of the crew,used to rowing boats with sliding seats. To their credit many of the rowers quickly mastered the new technique and began to make rowing a trieres look the most natural thing in the world. The larger members of the crew, however, found that there was not enough room for them to row properly (Fig.23). If they rowed with their legs straight they overlapped their seats at the back and obstructed the forward swing of the rowers behind them. If they had their arms straight at the catch, like the rowers depicted in the Lenormant relief (Figure 1), they were not able to swing far enough forward and again obstructed the rowers behind them (Fig.24). Thus they were forced to row with arms and legs bent, unable to work with their maximum efficiency.

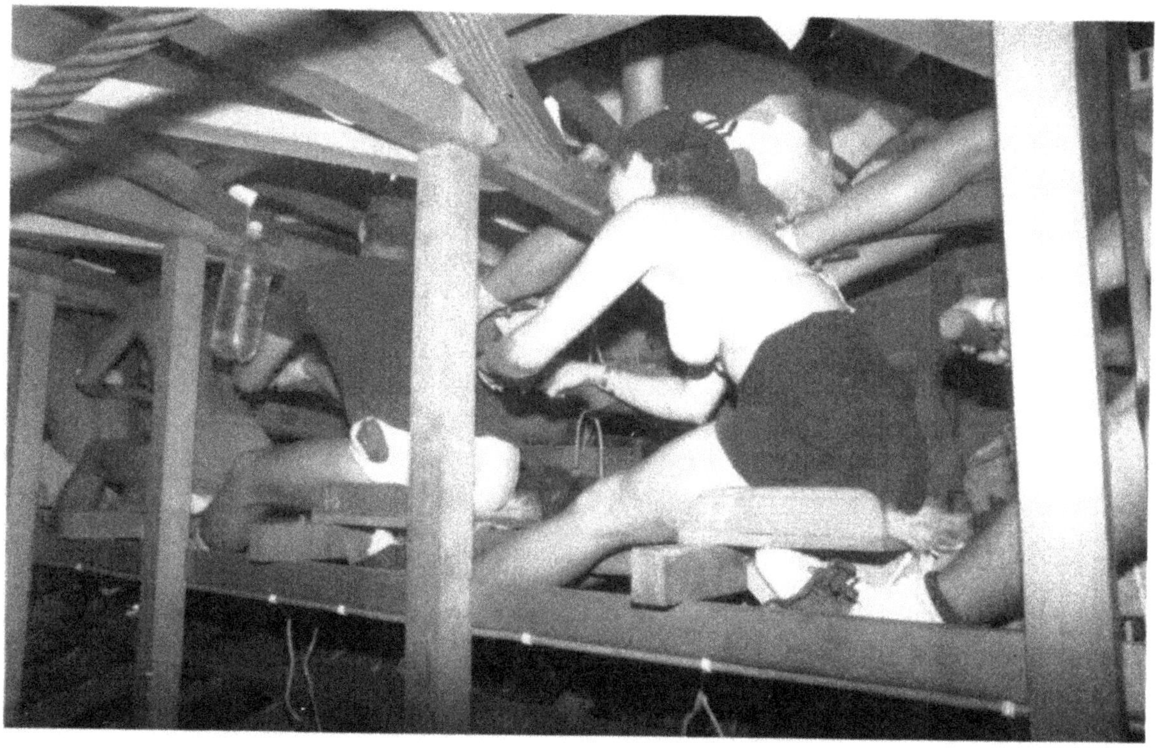

Figure 21 Thalamians at the Catch

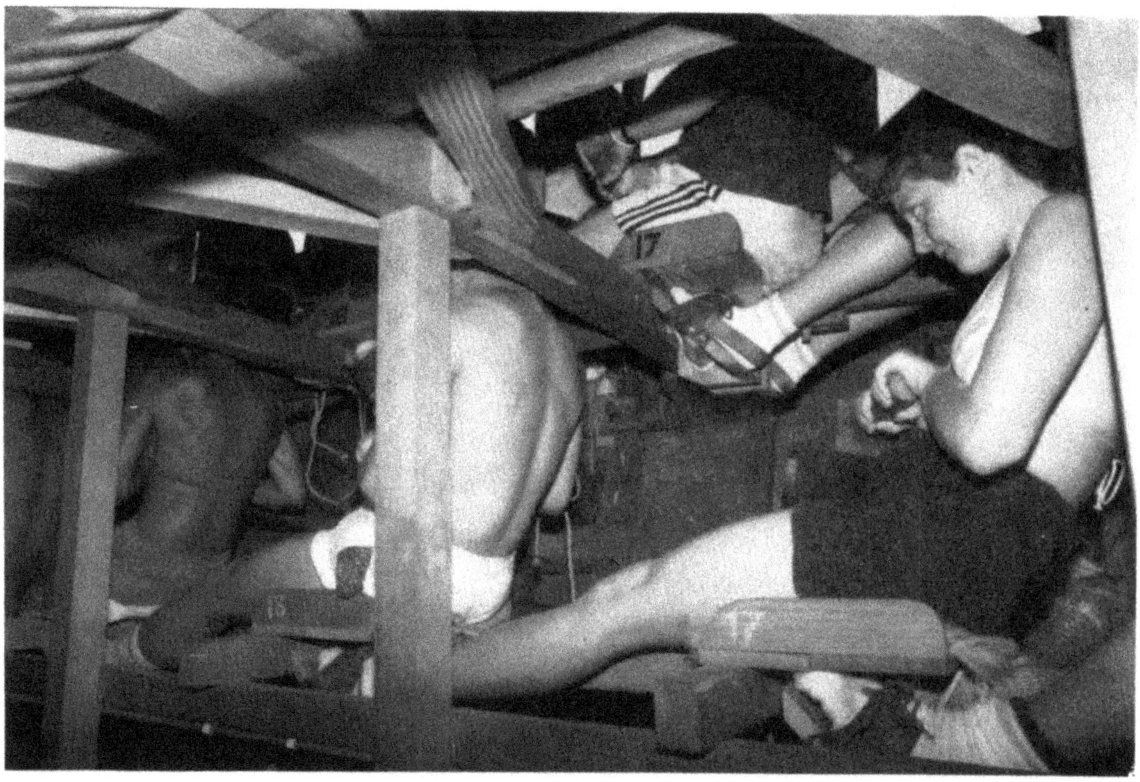

Figure 22 Thalamians at the Finish

CHAPTER THREE : THE SEA TRIALS

The rowers were best able to swing their bodies if they were well matched for size with those rowing forward and aft of them. It is certain that the oarsmen would have been on average shorter in stature than a modern crew of racing oarsmen. The women rowers aboard *Olympias* were probably quite close in height to Athenian men of the fifth century BC. It is perhaps significant that the best women rowers displayed very good technique indeed and achieved the full handle stroke of 0.8 m.(Figs.21 and 22).

The questionnaire revealed that rowers felt that a period of practice was required to get everyone rowing in the same style and swinging their bodies together. Since the zygians and thalamians could not see out of the ship, they had to follow the swing of the rower in front of them to keep time. Small errors of timimg thus tended to become magnified as they passed along the long files of rowers, and it was not unusual for the bow to be half a stroke behind the stern section (Figs.12 and 25).

Some rowers thought that the rating should be called out continuously to prevent errors building up along the ship. Although rowers at all three levels had similar work to do as far as all pulled an oar and needed to maintain a stroke and timing which harmonised both with their file (horizontally) and with their triad (vertically), there were also differences. One rower estimated that the effort required to pull an oar at the three levels was in the ratio of 5 (thranite): 4 (zygian):3 (thalamian).

Thranite Thucydides says (6.31) that the thranite oarsmen preparing for the Sicilian expedition were offered bonuses by their trierarchs. In *Olympias* the thranite position was physically the most demanding of the three. Even though the biggest and strongest rowers occupied the thranite level they still found it very hard work. The steep angle of the oar made pulling it awkward. At the finish hands were high, making it difficult to apply the downward force necessary to recover the blade (Fig.26). Rowers tended to change their inboard hand from an overhand to an underhand grip so as to spread the load among the different muscle groups. Such a hold by the thranite oarsmen is in fact depicted in a trieres on Trajan's Column in Rome (*AT Fig.4*).

When rowing overhand the rower found that his inboard elbow stuck out awkwardly at the finish. The counterweight block of the oar loom fouled the outboard wrist at the catch. One or two thranites complained that the loom of the zygian oar below had removed the skin from the back of their inboard leg. The thranite position had compensating advantages. Unlike the oarsmen at the other levels thranites were able to see the blades of their oars in the water.They could also observe what was going on about them and benefit from the cooling effect of any breeze that might be blowing. Most importantly of all perhaps they gained the impression that they had the most room in which to row, a mistaken impression in fact since all the oarsmen were spaced equally along the ship.

Zygian Although the zygian position was not physically the easiest to row in, it was easiest in terms of technique. Admittedly they could not see the blades of their oars in the water, but they were not hampered by apparent lack of head space as the thalamians were, and the angle of their oars was less awkward than that of the thranites'. Moreover in most of the seats at the zygian level oarcrew were relatively safe from the danger of 'catching a crab', unlike the thalamians. Bow zygians at positions 3,4,and 5 could and occasionally did have trouble with 'crabs' since the space for their furniture was reduced. It was possible for their blades to be jammed

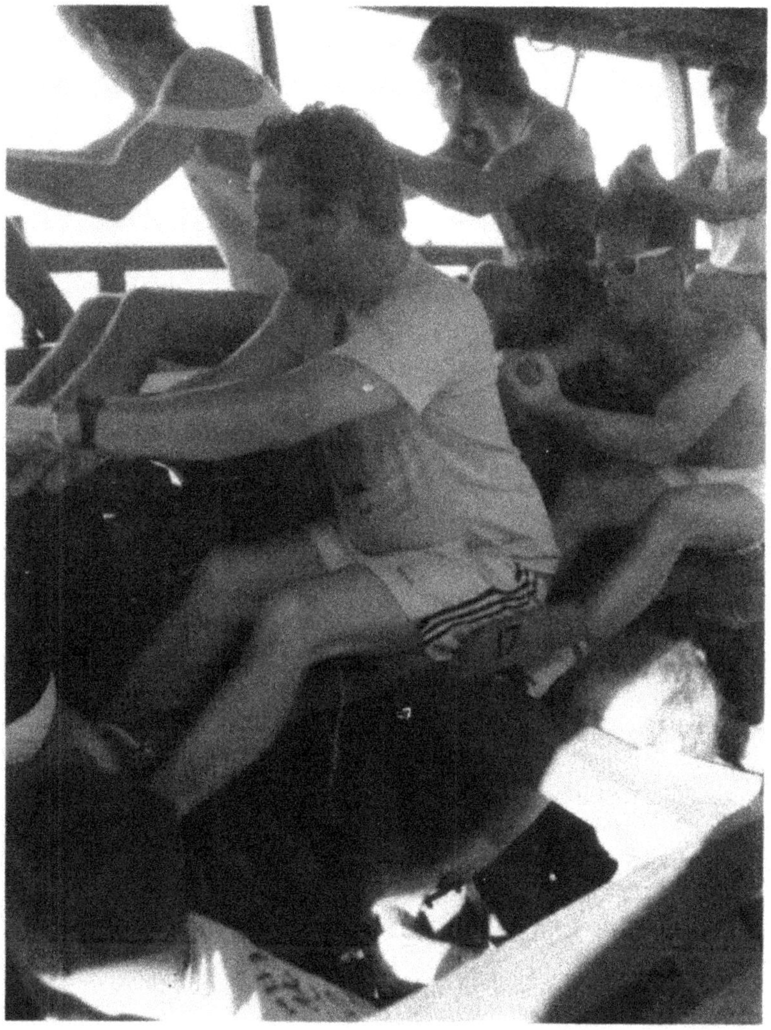

Figure 23

Oversized Oarcrew

(Zygian No.17 Port)

Figure 24 Oversized Oarcrew (Zygian No.12 Starboard)

against the thranite stretcher support and the stretcher itself. The less experienced rowers were given the zygian level, but it should not be thought that anyone however inexperienced or weak could be set to row there. Zygians still needed strength and rowing skills if they were to avoid endangering the safety of the thalamians below them and work efficiently within their triads.

Thalamian The thalamian position was the least glamorous in the ship. It was cramped, airless, smelly and potentially dangerous. That it was physically the least demanding of the three levels scarcely seemed to compensate for its disadvantages. However the thalamians had a vital role to play in the trials since they set the length of the stroke. Achievement by them of the longest possible stroke would be the main factor enabling *Olympias* to reach her maximum possible speed.

Thalamian stroke length was limited by the transverse beams of the ship's internal structure. The movement of the thalamian's head during the stroke was determined by the beams in front of and behind him. If the maximum stroke length was to be obtained then at the catch the rower was touching the beam in front of him with his face and the beam behind him with the back of his head. Maximum handle stroke was about 0.8 metres (nearly the distance between one tholepin and the next in a file of oarsmen (Fig.9). This maximum could only be obtained by a few groups of rowers who were homogeneous in stature, confident and skilled. Large rowers at the thalamian level obstructed the body swing of the others in their file (Fig.27). Thalamians needed confidence to leave their blades in the water into the finish of their stroke since it was possible for them to be forced into 'catching a crab' by a clumsy zygian above them. Ford Weiskittel provides a subjective view of this experience in Regatta of November 1987.

Paul Lipke writes:' It was not necessary for a zygian actually to trap a thalamian blade to cause the latter to 'crab', though this was by far the most common cause until the practice of separating these levels with a thranite oar was instituted after outing 4. Particularly in early training it did not take much for even a relatively skilled zygian to knock accidentally into his thalamian's blade during recovery. If the strike was sufficiently strong to cause the lower edge of the thalamian's blade to hit the water, and/or throw the thalamian out of step, water pressure would do the rest with a result identical to that described. The only exception occurred with particularly fast, strong and/or skilled thalamians who could somehow manage to recover. In fairness to the zygians, many thalamians were initially and understandably unwilling to swing their torsos much and risk hitting their heads on the transverse beams. As a consequence they rowed a very short stroke using their upper arm strength. This meant that adjacent thalamians who tried to reach fully and properly using their upper torso inevitably had to stop short or hit their neighbour in the kidneys. It also meant that the zygians, not restricted by transverse beams, were hard pressed not to overreach even after they became conscious of the problem. It only took a moment's loss of concentration (easy enough during several hours of rowing in intense heat) to get out of step.

After solving the problem of how to prevent zygian overreaching, 'crabbing' was reduced to approximately one an hour. Two veterans then tried stroke-lengthening technique with considerable success, though there was insufficient time to experiment and train with it enough to judge its full (apparently substantial) potential. To get thalamians to extend their stroke through increased torso swing the zygians were stopped rowing

Figure 25 Bow Oars Lagging Behind the Stroke

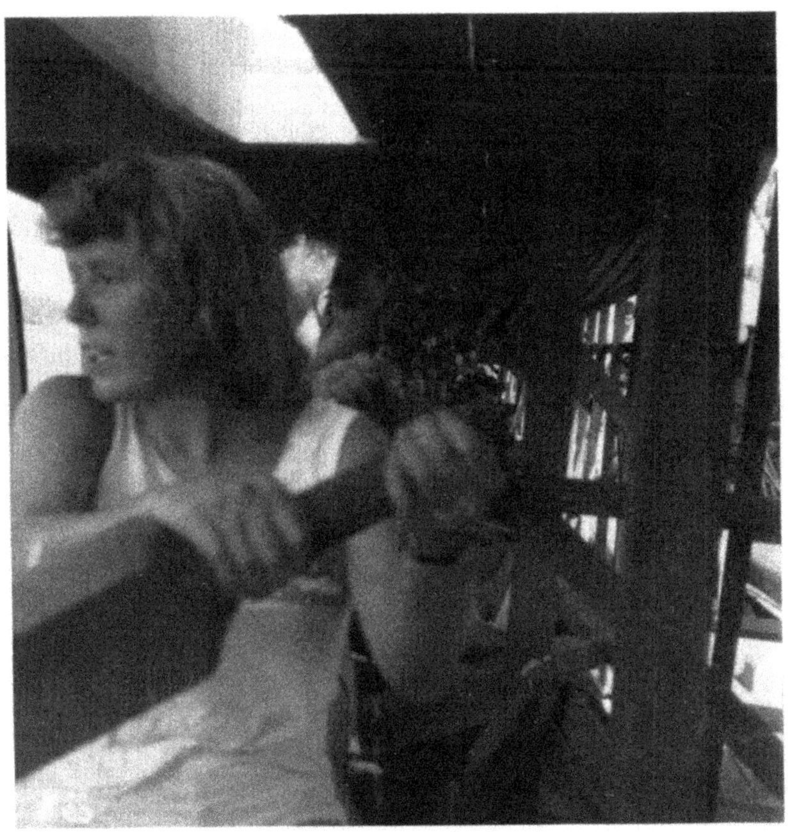

Figure 26 Thranite Finish Position

so that they could reach down and pad the transverse beam with a hand. The practice appeared to help thalamians to overcome their fear of the beam while also allowing them opportunity to feel just where the optimum stroke length occurred according to their individual physique.

As evidence of the importance of correct rowing, several recorders and team leaders noted independently that three well matched, excellent, adjacent, thalamians not only enabled their six other triad members to row properly with longer, stronger strokes, but the adjoining triads fore and aft suffered less interference. The three affected twelve others. Conversely, it showed that one poor thalamian could ruin his triad and both adjacent ones, a ratio of 1:8. Thus the single weakest rower can do more to affect performance than the single strongest one.

Some team coaches also tried stopping a dozen rowers round a trapped thalamian while the ship maintained full way with the remaining oarsmen pulling, but under full way the offending oar could not always be moved, even with the zygian above shoving hard down with his heel. The alternative was to stop the whole crew with a sharp blast of the whistle. Concurrently the thalamians were taught a procedure when caught in an impending 'crab'. By leaning their upper torso outboard into the open triangle formed by the side of the hull and the oar-loom they were able to allow the uncontrolled loom to come forward against the beam without trapping them in between.

3.2.3.3. *Speed under Oar*

A vital factor in the long naval supremacy of the trieres was her great speed. Xenophon gives a contemporary view of the importance of that factor in *Household Management* (8.8) : why is the trieres, fully manned, such a terror to the enemy and a joy to her friends except by reason of her speed through the water?

In battle the trieres needed to sprint short distances to gain a tactical advantage over her opponents with a view to ramming attacks (for examples see *AT pp.68-78*). She was also used to undertake long voyages at speed (for examples see *AT pp.94-106*).The aim of the trials was to conduct a series of tests designed to monitor the progress of the crew and compare the performance of *Olympias* with the performance of fifth-century triereis as shown in the historical sources. The speed of *Olympias* was to be measured over short bursts, measured distances of 500,1000, and 2000 metres, and eventually on longer voyages of up to 24 hours (including stops). It seemed unlikely that in the first set of trials the crew would make sufficient progress to undertake a long passage, and this proved to be the case. Measurement of speed on a passage of 24 hours or more must await the acquisition of greater experience and skill in rowing the trieres.

Equipment and Measurement For the measurement of speed during short bursts and during sustained periods of rowing the very simple "Dutchman's log" method was used. The forward outboard corner of the canopy-deck and the upper rudder bearing aft(on the port or starboard side depending on the direction of the wind) were chosen as convenient points from which to observe passage past a floating object. The distance between these two points was measured with a tape parallel to the centre line of the ship. The distance was 92,5 feet.

To obtain a measurement of the ship's speed one member of the recording team stood at the forward point armed with a good supply of wood chips. A second recorder, with a stopwatch, stood by the after point. The first recorder then, with one arm raised, threw a chip into the water with the

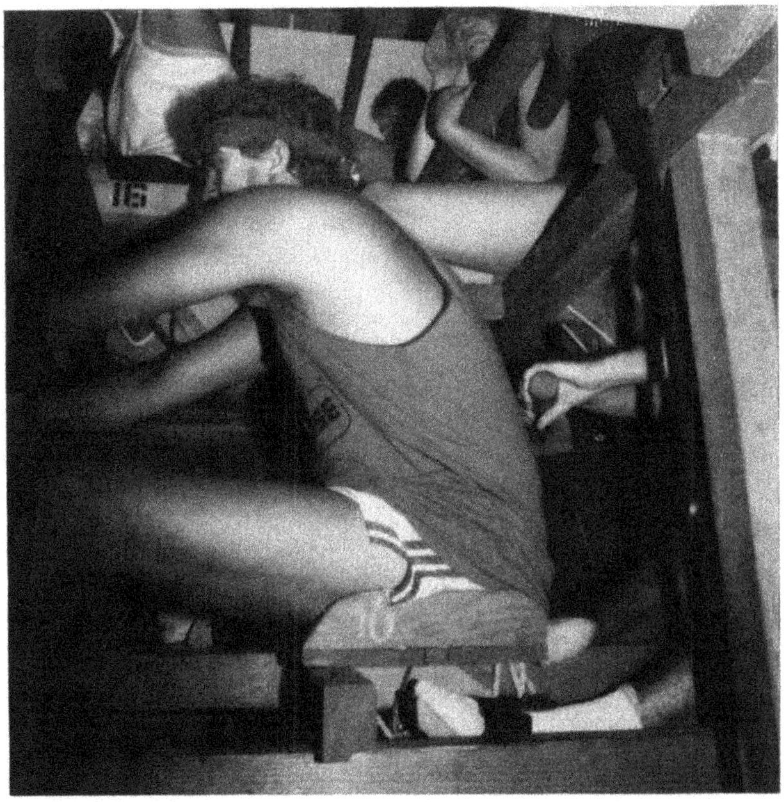

Figure 27 Oversized Thalamian

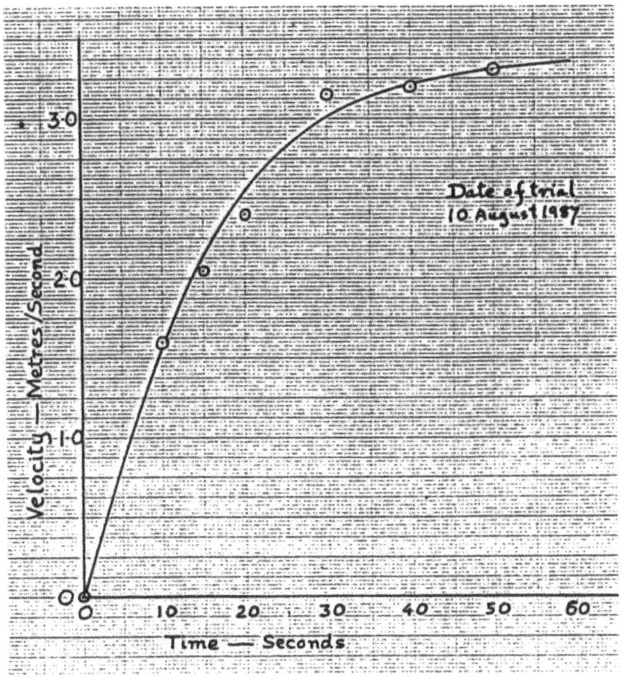

Figure 28 Velocity - Time: during Acceleration Trial

other so that it fell well forward of the canopy-deck and far enough out to avoid the turbulence caused by the oars. When the forward recorder came level with the chip he had just thrown he would drop his arm and the recorder aft would start his watch. When the aft recorder came level with the wood chip he stopped his watch. The speed of the ship in feet/seconds = 92.5 ÷ t. The speed in knots = (92.5 ÷ t) x (3600 ÷ 6080) where 3600 seconds = 1 hour and a nautical mile is by convention 6080 feet.

For the measurement of *Olympias*'s speed over distances of 500.1000 and 5000 metres it was necessary to prepare a course for the trieres to follow in Poros bay. This was done by means of a large scale Greek Government chart supplied by Captain Papadas. The course ran parallel to the coast of Poros island well clear of the ferry lanes : 296 °Magnetic or its reciprocal 116°Magnetic. Five landmarks were selected,all,to reduce the margin of error, at right-angles to *Olympias*'s course.

They were as follows (see Fig.35):-

Starting point	: an obelisk on the point below Poros Hotel
500 metres	: house with round arches at most northerly part of the bay of Neorion
1000 metres	: red rocks just west of the bay beyond Neorion
1500 metres	: tip of rocky promontory at S.E.end of Daskalio island
2000 metres	: bluff on w.side of a small inlet,the last before the light on the rocks at the exit of Poros Bay

When a mark was bearing 26° Magnetic, the designated point of the measured course had been reached.

When it was intended to measure speed over a fixed distance *Olympias* would be set on the correct course with a compass. The Helmsman would then choose a suitable mark to steer for. A recorder with a hand bearing compass was placed in the stern close to another recorder with a stopwatch. When *Olympias* reached the first mark the recorder with the compass called out to the recorder with the stopwatch who then started the stopwatch. Intermediate distances were also called out by the recorder with the compass and the elapsed time noted by the recorder with the stopwatch. On the completion of the chosen distance the watch was stopped. All fixed distance measurements were made with a running start. It was not always possible to carry out a reciprocal run (i.e. one in the opposite direction), but since there was negligible current in the bay and most measurements were taken under practically windless conditions, this was not thought necessary.

When speed measurements were being taken rating watches were used to record the number of strokes per minute (s.p.m.). The amount of effort required of the rowers by the Rowing Master was also noted down. Wind speed and direction were estimated by the experienced seamen of the team and recorded (Annex 5).An analysis by Timothy Shaw, the British Helmsman, of the power and pace achieved at Poros is appended at Annex 7.

Rowing with all oars On the very first outing, although the oarcrew were quite unused to the ship, it was possible to pull all the oars together and reach a speed of 4.2 knots, admittedly with a Force 3 wind astern. As one would expect, the rowers adapted to the different technique required and became fitter and stronger as the trials progressed. After outing 11 five tonnes of ballast were taken aboard to bring the waterline to the desired level (see above p.20). The rowers would thus, it was hoped, be able to use their oars to better effect and make the ship go faster.

Speeds attained at *light pressure* ranged from 3.04 knots on outing 2 with a Force 2 head wind to 6.09 knots on outing 17 with a Force 3 stern wind. The lowest rating recorded at light pressure was 25 s.p.m. on outing 7, the highest at light pressure was 34 s.p.m. on outing 17. Further analysis was carried out to obtain average (i.e., arithmetic mean) and median (i.e. arithmetic mean of the fastest and slowest recorded) speeds under different conditions.

	Average speed	Median speed
Before ballasting	4.73 knots	4.97 knots
After ballasting	4.89 knots	4.88 knots
Overall	4.79 knots	4.91 knots
With no wind	4.95 knots	4.97 knots

The median rating for light pressure was 30 s.p.m.

The ship completed one 2000 metre run at light pressure on outing 9 at a rating of 29.5-30 s.p.m. rising to 31 s.p.m. towards the end of the run. There was no wind. The times recorded at the various marks were as follows:-

		Between marks	Speed,knots,average between marks
0 metres	0 minutes 0 seconds		
500 metres	3 minutes 10 seconds	3 min.10 sec.	5.11
1000 metres	6 minutes 33 seconds	3 min.23 sec.	4.78
1500 metres	9 minutes 38 seconds	3 min.5 sec.	5.25
2000 metres	12 minutes 55 seconds	3 min.17 sec.	4.93

Average speed over the 2000 metres was 5.01 knots, which may be compared with the figures of 5.09, 5.36, 5.12, 5.11, 5.08 and 5,55 (average 5.22)knots obtained by the Dutchman's log during this same run.The longest period of continuous rowing at light pressure with all oars in operation was 15 minutes on outing 20.(See Annex 5, Table 1 for the full list of light pressure measurements).

Only two measurements of speed were obtained at *half pressure*, 4.63 and 4.73 knots on outing 19,at a rating of 29 s.p.m. into a Force 3 head wind. This was the only half pressure run of the trials. *Three-quarter pressure* runs were also rare. On outing 18 the ship completed one 1000 metre run at three-quarter pressure at a rating of 33 s.p.m.. There was no wind.

The times recorded were as follows: -

500 metres	2 minutes 55 seconds
1000 metres	5 minutes 43 seconds

Average speed over the 1000 metre course was 5.67 knots, a little faster than the speeds of 5.37 and 5.26 knots obtained by the Dutchmazn's log during this run. This was the longest period of continuous rowing at three-quarters pressure with all oars in operation, in total just over six minutes (See also Annex 5 Table 2).

At *firm pressure* a good number of speed measurements were made both from very short bursts of 10-15 strokes during the practice of acceleration, and from longer runs over the 500 and 1000 metre distances. Speeds attained at firm pressure ranged from 5.65 knots on outing 6 into a Force 3 head wind to 6.95 knots on outing 7 into a Force 1 head wind. The lowest rating recorded at firm pressure was 29 s.p.m. on outing 7. The highest was on outing 11 when it touched 44 s.p.m..

Further analysis was carried out to obtain mean and median speeds under different conditions:

	Average Speed	Mean Speed
Before ballasting	6.20 kn.	6.13 kn.
After ballasting	6.03 kn.	6.09 kn.
Overall	6.09 kn.	6.09 kn.
With no wind	6.18 kn.	6.09 kn.

The median rating for firm pressure runs was 38 s.p.m.

The ship completed five 500 metre runs and one 1000 metre run at firm pressure. Three of the 500 metre runs were rowed on outing 16.

Run 1 Rating 38-39 s.p.m. No wind.
500 metres in 2 minutes 26 seconds
Average speed over the course was 6.65 knots. No speeds were obtained by Dutchman's log.

Run 2 Rating 39 s.p.m. No wind. An emergency 'easy oar' at the end of this run meant that the effort was not sustained past the mark but a time was obtained notwithstanding.
500 metres in 2 minutes 45 seconds
Average speed over the course was 5.89 kn. Three speed readings obtained with the Dutchman's log were all of 6.09 kn..

Run 3 Rating 39 s.p.m. No wind
500 metres in 2 minutes 31 seconds
Average speed over the course was 6.44 kn.,slightly higher than the speeds obtained with the Dutchman's log: 6.02,6.41 and 6.09 kn.

The other 500 metre runs at firm pressure were taken on outing 21.

Run 1 Rating 36-38 s.p.m. Force 2 head wind
500 metres in 2 minutes 30 seconds
Average speed over the course was 6,47 kn. The Dutchman's log gave speeds of 6.00 and 6.14 kn.

Run 2 Rating 38 s.p.m. Force 2 head wind
500 metres in 2 minutes 30 seconds
Average speed over the course was 6.47 knots. The Dutchman's log gave speeds of 5.86,6.02,6.11,6.06 and 6.28 knots. Again the speed given by the measured distance was higher than the speeds given by the Dutchman's log.

The 1000 metre run at firm pressure was rowed on outing 18.
Rating 32-34 s.p.m. No wind
500 metres in 2 minutes 38 seconds
1000 metres in 5 minutes 26 seconds
Average speed over the full course was 5,96 knots. This was the longest piece of continuous rowing recorded at firm pressure with all the oars in operation,in total just under six minutes. The down-draught from the helicopter carrying a BBC film crew made it impossible to carry out the Dutchman's log procedure (see also Annex 5 Table 3).

The ship's top speed, measured during a short burst by the Dutchman's log was 6.95 knots. High ratings produced the highest speeds with 38 s.p.m. which was the norm for high pressure runs. The maximum rating reached was 44 s.p.m.. The oarcrew could sustain 6.5 kn. over 500 m. and 6.00 kn. over 1000 m.. A normal light pressure speed was about 5 kn. although with a moderate following wind the ship reached 6kn.. Speeds were noticeably affected by the wind, but since nothing stronger than a moderate breeze was experienced on the trials there is still much to find out about *Olympias*'s performance under windy conditions. The longest recorded period of continuous rowing at firm pressure was 6 minutes, at light pressure 15 minutes. On outing 21 there was a period of continuous rowing of about 25 minutes, but it was unrecorded since the recording team were at the oar. Ballasting produced greater convenience for the oarsmen but had no marked effect on speed. If anything the ship was slower after ballasting.

Speeds obtained with all the oars were disappointing.A top speed of 6.95 knots during a short burst corresponds to what a trieres in ancient times might expect to achieve while cruising for a 'long day' of 10-12 hours at the oar. Heavy, ill-balanced oars and rowers who were ill-assorted in size, strength, experience and skill and thus unable to get the best out of the ship, both contributed to a performance which by comparison with antiquity was unimpressive. Yet, ill-assorted though they were, the oarcrew has one achievement to their lasting credit; they demonstrated by their enthusiasm and intelligent determination not only that *Olympias* could be rowed, but that she could be rowed very beautifully.

Backing Water Going astern was an important manoeuvre in ancient naval warfare, in which it followed, and was the reciprocal action in, the action of ramming. It also enabled the helmsman to keep out of the melee until he had selected a suitable target. Moreover it was vital when ships in line abreast were withdrawing in good order keeping their rams pointed towards an advancing enemy (see Herodotus 8.84 and Thucydides 3.78 and 7.70.4.). In oared ships from Homer onwards backing water was the routine manoeuvre which preceded mooring, with anchors dropped from the bows and stern lines run ashore. At Poros it was accordingly recognised as necessary to practise this manoeuvre with *Olympias*, and some data were collected about it(Annex 5 Table 7).

The oarcrew as a whole never really mastered the art of backing water. The thranites were used on their own regularly to bring the ship stern first to her moorings. The thranites and zygians also were able to back her quite successfully; and on outing 13 reached a speed of 2.59 kn.into a Force 1 wind going astern. However when the whole crew attempted this manoeuvre, the blades began to clash. Since there was no time to perfect the technique of backing water with a whole crew, thranites and zygians only were used for it in the trials at Poros.

3.2.3.4.*Acceleration*

A trieres able to accelerate faster than her opponents would have had a great advantage in battle. The ability to make rapid use of a gap in the enemy's line was the essence of the *diekplous* (See above p.9 and *AT* Index under' Breakthrough'). Triereis stationed in the middle of a defensive circle were so placed as to be able to move quickly to any point in the perimeter where a breakthrough was being effected (Thuc.2.83). Historically therefore acceleration was an important quality in an oared ship built for ramming and merited investigation in the case of the *Olympias*.

Measurement A short course was laid out along the quay of the Petty Officers'School at Poros by means of a tape measure and a hand bearing compass. It followed the line of the quay wall and was 500 ft long. The course was on a bearing of 329°Magnetic. "Coke" cans were placed at 0 ft,50 ft,100 ft,150 ft,200 ft,300 ft,400 ft and 500 ft. The compass was then used to set up back markers in the shape of more cans behind, and at 90°to the line of the quay wall markers to provide a series of transits. Each back marker was at a bearing of 59° Magnetic from its quay wall counterpart. These improvised markers were highly visible, perhaps too visible, since four of them were cleared away as rubbish. Thereafter a guard was placed on them to prevent their removal.

When an acceleration run was to take place the ship was rowed into position 50 m. or so offshore and not quite in line with the markers at 0 ft. She was then lined up on a course parallel to the quay wall markers, that is 329°, using a compass. One recorder then stood with his heels touching the starboard upper rudder bearing with his arm raised. The ship was slowly manoeuvred on to the starting point of the course.

When the 0 ft markers appeared in line to the recorder standing by the rudder bearing, he dropped his arm. This acted as a signal to the Rowing Master to set the crew rowing and for another recorder to start a stopwatch. The recorder by the rudder bearing raised his arm again, and dropped it when the 50 ft markers were in line, and so on until the end of the course had been reached. The stopwatch was set to store the interval times in its memory while continuing to run on. Once the run had been completed the times at the various distances could be recalled from the watch's memory.

Acceleration from a standing start was practised a number of times in open water to discover the best and safest way of getting the ship moving quickly. Five strokes at light pressure followed by ten at three-quarter pressure, fifteen at firm pressure and then 'easy all' was the sequence adopted. Winding down(see Glossary) after firm pressure was abandoned as too dangerous since, with ratings rising to 44 s.p.m. during these runs, it was easy to lose control of the blade on the wind down.

Three runs were attempted over the acceleration course. The first two, on outing 12, were futile because some of the markers had been removed. The third run took place on outing 17. There was a Force 3 beam wind which made the passage near the two steel buoys anchored off the quay somewhat hazardous. Our Helmsman declined to take responsibility for the ship under these circumstances and handed command back to the Captain. The ship was steered by the second officer during this run. The oarcrew rowed five strokes at light pressure, ten strokes at three-quarter pressure, fifteen strokes at firm pressure and then went 'easy all'. The rating started at 26 s.p.m. and rose to 42 s.p.m.:

Times at the markers	: 50 feet	15.5 seconds
	100 feet	22.5 seconds
	150 feet	28.4 seconds
	200 feet	32.8 seconds
	300 feet	42.6 seconds
	400 feet	52.1 seconds
	500 feet	63.7 seconds

The crew was at easy between the 400 and 500 ft markers.
Distance (in metres) was plotted against time to give a series of velocity values by differentiation.

Time (secs)	Velocity (m./sec)
0	0
10	1.6
15	2.05
20	2.4
30	3.15
40	3.20
50	3.3

Velocity was then plotted against time to give a series of acceleration values (see Fig.28) by differentiating again.

Time (secs)	Acceleration (m./secs)
0	0.18
10	0.12
20	0.11
30	0.04
40	0.02
50	0.01

The inaccuracy of this method is recognised, but no other was available in the circumstances.

The ship reached a speed of six knots after 30 seconds. It might be possible to pull her away faster if the earlier strokes at light pressure were replaced by strokes at three-quarter pressure, but this procedure might involve risk of injury to the rowers. Lighter oars would of course enable the crew to accelerate the ship more quickly. The rate achieved was however very satisfactory.

Tim Watson plotted the graphs and calculated the velocity and acceleration values. In Annex 7 Timothy Shaw derives consistent values from consideration of oar geometry, slip, power and the resistance curve of the ship. The latter (Fig.37) is derived from model hull-towing experiments by the National Technical University of Athens (Report No.NAL 06-F-1985 by Professor T.Loukakis).

3.2.3.5. *Turning* Although the chief characteristic of the trieres was speed, she also needed to be highly manoeuvrable in battle. Fast turning towards or away from an opponent enabled the commander to choose the best tactical position in attack or defence. *Diekploi* 'breakthroughs' and *anastrophai* 'quick turns' are spoken of (Thuc.2.89.8) as the proper tactics of ships that move faster in the water. Unfortunately it is not possible to draw conclusions about the precise degree of speed or sharpness with which a fifth-century Athenian trieres was able to turn. But the turning effectiveness of *Olympias* was clearly one of the characteristics which had to be explored as fully as possible in the Poros trials. Since under normal circumstances a trieres would have had to make all sorts of turns, at high speed, when stationary, or even when going astern, *Olympias*'s turning ability was tested in a variety of ways to attempt to simulate different possible battle situations.

Measurement Turns were measured by means of a compass placed on the canopy deck. A recorder with a stopwatch sat by the compass and started the watch when the helmsman applied the rudders. In the case of turns executed solely with the oars the watch was started when the Rowing Master gave the command to row. The time taken for the ship's head to swing by 90° was noted and the watch was stopped when the ship's head had swung by 180°. Rating watches were used to record the number of s.p.m.; and the amount of effort asked of the rowers was also noted. Wind speed and direction was estimated by the experienced seamen of the team and recorded. A very rough estimate of the distance travelled by the ship while making the turn was also made and noted.

In the ship's cruises round Poros bay she was frequently to be turned. The turn was most easily accomplished by applying the rudders while the oarcrew continued to row at light pressure. The rhythm of the stroke was thus continued undisturbed. This operation could be called a cruising turn. A typical cruising turn was executed and recorded on Outing 7.

Figure 29 Helm hard-over

Figure 30

Turning to Starboard:

Starboard Oars at 'easy'

There was no wind. Full helm was applied as the whole crew continued rowing at light pressure, with a rating of 28 s.p.m.. The turn of 90° took 37 seconds, while the full turn of 180° took 1 minute 18 seconds.It was estimated that the first 90° turn was effected within 2½ lengths and the second 90° turn in a further 2 lengths. A tighter turn was possible by holding water with the oars on the inside of the turn, but this could only safely be done if the ship was not moving too fast, e.g., when the crew had only taken ten strokes at light pressure from a standing start. The Rowing Master considered it too dangerous to attempt to hold water on any of the firm pressure turns.

Another manoeuvre tried was a stationary turn of the ship, exactly analogous to the 'spinning of the boat' manoeuvre performed by eights and fours except that, because of the difficulties with backing, only the thranites and zygians were used. The stationary turn was only tried once, on outing 10. The 90° took 58 seconds and the 180° exactly two minutes. The stationary turn is rather slow and is only likely to have been used, e.g., in moving from the stationary line abreast formation into line ahead or in a small harbour.

Turning the ship while she was at maximum speed, simulating the sort of turn which might have been carried out in battle, proved to be an exciting manoeuvre which was popular with the oarcrew. A number of these battle turns was attempted as we experimented to find out which was the quickest way of turning the ship. First, on outing 10, a turn to port was tried with all oars at firm pressure and a Force 3 wind from astern. Full helm was applied. The 90° turn took 35 seconds, the full 180° turn 1 minute 7 seconds. The ship heeled appreciably to port during this turn making it difficult for the port side oars to maintain their rhythm. Accordingly, when subsequent battle turns were effected, the inside oars pulled five strokes only after the helm had been applied before going easy, or on occasion went easy as soon as the helm was applied. The fastest 180° turn was made on outing 10 with a Force 3 wind astern. The ship went into a turn to starboard with all oars at firm pressure. Full helm was applied (Fig.29) and the starboard oars continued to row for five strokes before going easy (Fig.30). Port oars rowed on at a rating of 33½ s.p.m.. The 90° turn took 30 seconds and the full 180° turn 1 minute 1 second. Each 90° was effected in about two ship's lengths. Several other turns matched the 30 seconds for the first 90°, including two when the inside oars went to easy immediately the helm was applied. Making a turn when the wind was pushing the bow round was easier than trying to turn up into the wind. A 90° turn to port into a Force 3 wind on outing 10, which took only 30 seconds represented a better result than any of the other turns.

As part of a simulated withdrawal after ramming, the ship executed some 90° turns while going astern. All these reverse turns were made from a standing start with thranites and zygians only. Since backing speeds were slow it was possible for the rowers on the inside of the turn to hold water while those on the outside backed water.

Timothy Shaw, who was Helmsman for most of the Poros trials, reports that the ship handled very well indeed, both under oar and under sail. To act most efficiently the two rudders had to be at the same angle. He had no problem in achieving this. When full helm was applied the helmsman needed an assistant to take one of the tillers since they were then too far apart for one man to operate both.

Figure 31 Sailing before the Wind on the Starboard Quarter

Outing No.	Sails Set	Force of Wind (Beaufort Scale) and Direction	Ship Speed V_s Knots	K.P.H.	Est. True Wind Speed V_T, K.P.H.	Sailing Efficiency V_s/V_T
1	Bare Poles	3 Following	2·0	3·6	13	0·28
4	Bare Poles	4 Following	2·3	4·1	18	0·23
15	Boat Sail	1 Starboard Quarter	1·0	1·8	5	0·36
15	Boat and Main	1 Starboard Quarter	2·3	4·2	5	0·85
17	Bare Poles	3 Following	2·0	3·6	13	0·28
19	Boat and Main	2-3 Starboard Quarter	5·1	9·3	11	0·84
19	Boat and Main	2-3 Starboard Quarter	5·9	10·7	13	0·82
20	Boat and Main	3 Port Beam	Est.>7	12·7	16	0·79

O.T.P.R. FEB 1988

Figure 32 Sailing Table

3.2.3.6. *Sailing*

Rig Handling Hoisting the small forward, boat, sail (see *AT* Fig.68) was easy since it was light. The main yard and sail were extremely heavy for the reasons given above (3.2.1.4). Both sails were hoisted brailed up. A slippery hitch was made to each brail at the fairlead so that a sharp pull was sufficient to unfurl the sail. Hauling on the tack and sheet helped the process.

Double halyards (see *AT* Fig.68) were hauled together but not married. The gangway aft of the mainmast was just wide and long enough for the two hauling gangs.

Positions for the sheet and tack leads were found after a little trial and error. The tack was tied off by a single slippery hitch close up to the lead. The sheet was led aft across the canopy-deck to the pin rail. Both main braces were led to the samson posts (see Glossary) just ahead of the helmsman. This arrangement permitted most of the mainsail's controls to be managed from an area near the helm.

The boat sail's controls were just aft of the boat mast; and a small deck crew under a petty officer trimmed the sail as ordered.

In the light winds experienced during the trials one man was sufficient for each brail, but in a breeze two or three men will be needed for each. The most practical way of furling proved to be to brail from the middle of the foot of the sail outwards as the sheet and tack were gradually eased.

First Sailing Period The sails were hoisted brailed before casting off from the quay and proceeding clear under oars. About a dozen rowers joined the deck party for this and subsequent sailing periods. All claimed some sailing experience, and those with the most soon showed an increasing understanding of the lost art of handling the ancient Mediterranean square sail rig.

The wind was on the starboard quarter and from the sea conditions was judged to be Force 1 (from 1.6 to 4.8 k.p.h.). All oarsmen were warned not to drop their blades into the water on this and later occasions when under sail because of the real danger of oars being swept aft.

The boat sail was unfurled and trimmed. Steerage was maintained positively even in such light airs, indicating how effective the boat sail would have been, used by itself, in ancient times. It was estimated that the ship made 1 knot under this rig in these conditions. How much the hull, with its high upcurving stern, contributed to this speed was not measured.

The mainsail was unfurled and trimmed. In the almost following wind the boat sail immediately went limp and even backed slightly. This effect was caused by the turbulent area on the lee side of the mainsail large enough to include the boat sail. From the behaviour of the boat sail it appeared that the air flowing over the main yard became a down draught over it. This unwanted effect was remedied by brailing up the middle part of the mainsail about 4 ft.(1.3 m.). The boat sail then filled correctly. With both sails drawing the ship made 2.34 knots through the water (Dutchman's log). In these light airs such progress showed how fast the ship was likely to be under sail.

Second Sailing Period This outing followed the same routine as the first. Slightly better sailing conditions were experienced. The wind was on the starboard quarter (Fig.31). From sea conditions it was judged to be Force 2 edging into Force 3,, i.e., about 13 k.p.h., and described as a light breeze. The speed of the ship, as measured, was 5.12 knots and later 5.88 knots. These results were excellent in the light conditions. Brailing the middle part of the mainsail not only gave the boat sail clear air but pushed out the tacks of the mainsail as had been hoped.

CHAPTER THREE : THE SEA TRIALS

Although the trial area was generous for rowing it was too restricted for uncertain manoeuvring of a strange vessel with an untrained crew. To proceed from running down the length of the area to coming on to a reach and then a close reach before attempting to tack would have put the ship too close to the shore. At five knots there would have been less than twelve minutes to complete this manoeuvre. Given a knowledge of the ship's sailing capabilities and an experienced deck crew it would have been possible to beat back to the quay. As it was the decision was taken to furl sail and return under oars against the wind.

Third Sailing Period The ship was sailed for part of the return from a beach on the north east side of the island. The ship was rowed clear of the land and then sails were set and braced round to a breeze from the port beam. From sea conditions, i.e., large wavelets with breaking crests infrequently seen, the wind was judged to be Force 3(8-12 knots,15-20 k.p.h.), a gentle breeze. The boat sail drew well with no backing. The heel was estimated to be not more than 5°, and the ship felt reassuringly stiff. Leeway could not be judged from the twin tracks of the rudders. By taking bearings it was found to be less than 7° from the ship's heading.

Gradually the wind drew to just ahead of the beam and the yards were braced over. The ship continued to steer well and make a long,straight wake, but the chance was taken to brail up the after part of the mainsail. In the gentle breeze the change seemed to make no difference to the hull/sail balance. The leeway angle increased marginally to 7° while the tiller remained in the slightly lee-helm position.

From the tips of the rudder blades two trailing vortices had developed and these snaked beneath the wake as far as could be seen. Since these had not been seen before they indicate that the sailing speeds attained that afternoon exceeded the fastest rowing speeds attained up to that date. They supported the general awareness among the rowers that the ship was indeed slipping along faster than she had under oar and with very little strain on the slim hull.

The sails were brailed as the ship reached the channel between the island and the mainland and the wind was contrary. For the row home the yards were lowered to reduce windage for the rowers' benefit.

Conclusion The ship was sailed for a total of 1¼ hours during which data were collected which are displayed in Fig.32. The final column indicates that her sailing efficiency ratio (see Glossary) is likely to exceed 0.8 in all sailable conditions where there is a free wind. It suggests that the estimated 7 knots speed on the last sailing trial may be a little conservative. The ship has the potential to sail her theoretical maximum speed of about 12 knots (21.7 k.p.h.). At a sailing efficiency ratio of 0.8,then, a free wind of nearly 27 k.p.h., upper Force 4, should suffice for such performance.

The ship's ability to produce useful speed in light breezes is not to be ignored in the analysis of such fast voyages as are on record in classical times.

The lightening of the yards and the removal of the shrouds will facilitate the re-discovery of the potential of the square rig used by the trieres. Handling of this ancient Mediterranean rig is a technique to be learnt and enjoyed. Its use on *Olympias* draws back the curtain for nautical archaeologists on a method of handling sails developed over c.3000 years and out of use for c.1500 years, and should lead to an insight, tantalisingly partial, into the problem which brought first refinement and then ultimate abandonment in favour of more efficient rigs.

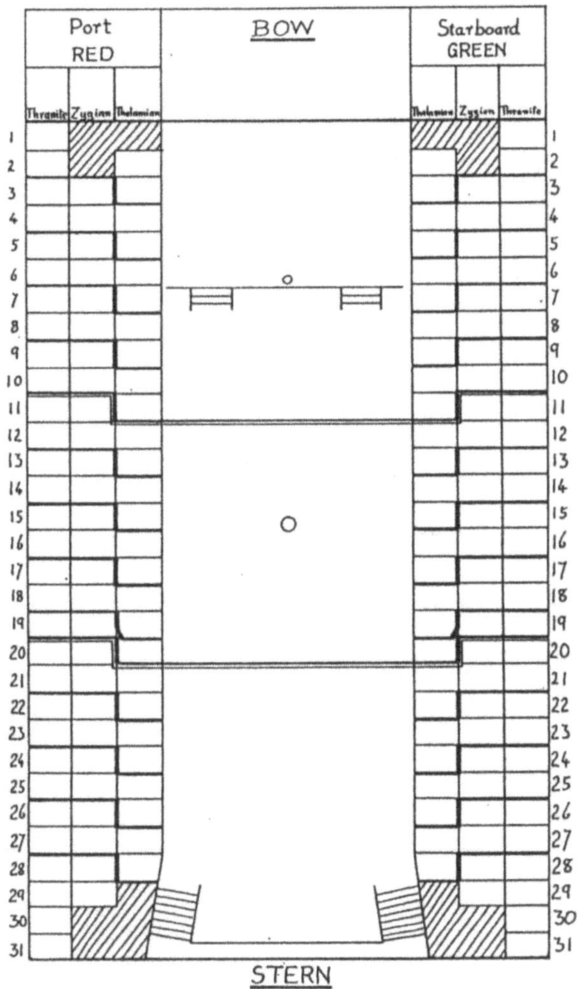

Figure 33 Crew Plan

Figure 34 Crew drawn up, ready to embark

3.2.4.*Embarkation, Mooring and Men on Deck (Epibatai)*

Embarkation Swift and disciplined embarkation, in the face of a surprise enemy move (cf.Thuc.2.89.9), was as necessary to the crew of a trieres as 'scrambling'is to the pilots of modern fighter aircraft. Disembarkation was less important except to the crews of triereis beachedin defeat. Emergency embarkations and disembarkations were not attempted at Poros since even with normal usage the ship's internal furniture often broke loose and further damage was feared. Nevertheless, it seemed worthwhile for the benefit of future crews to record the routine procedures which the crew observed when going aboard and ashore.

First a crew plan was chalked out on the quay in the form of a grid (Fig.33), large enough for rowers to stand in the various boxes which represented the oar positions (Fig,34). When the crew was mustered on the grid the team leaders shuffled their team members around into a satisfactory arrangement, when all the boxes were filled. Then the rowers led off, the bow sections leading. Port and starboard thranites moved up the single plank brow alternately on to the port quarter. Once aboard they descended the steps in the stern on to the gangway which ran down the middle of the ship, and made their way along it to their seats where they ran out their oars. Thranites embarked first because their oars when shipped not only hampered the passage of rowers along the gangway but also passed over the zygian seats. Once the thranite oars had been run out and were out of the way, the zygians whose oars when shipped also obstructed the gangway, were able to follow on. Finally the thalamians, the file closest to the gangway, boarded, took their seats and ran out their oars. On outing 1 embarkation took 21 minutes, but by outing 3 the time had dropped to 12 minutes, and did not alter significantly after that.

Provision of the second brow leading over the starboard quarter would have speeded up the embarkation a good deal, and since two *klimakides*, ladder brows, are among the gear for triereis listed in the naval inventories, it is reasonable to suppose that both were normally used in antiquity. The system described remained the same throughout the trials. Graeme Fife, who was one of *Olympias*'s crew at the Commissioning Ceremony, reports that a different procedure was employed by that crew. Rowers took their seats in bow and stern first. His impression was that they were able to board the ship faster than the British crew. At the ceremony they embarked at the double.

On campaign, triereis moored stern to on suitable beaches, and disembarked their crew down two ladder brows put out over the stern. On outing 20, when the ship made a day voyage and was moored stern to a beach for the midday meal, there was an opportunity to try this procedure. Once *Olympias* had been moored to the beach, port and starboard ladders were put out and the crew was successfully disembarked and re-embarked. Re-embarkation took 15 minutes. There is no doubt that in an emergency a trained and agile crew should be able to embark in a much shorter time.

Embarkation and disembarkation caused the crew no problem at Poros. There is however plenty of room for further experiment. With two ladders rigged and the ship's internal furniture fastened more securely, it would be interesting to try emergenmcy embarkation from beaches by a variety of different procedures.

Mooring Homer provides a detailed description of the operation of mooring stern first to the shore in *Iliad* 1.435-6: 'And rowing forward with the oars', i.e., pushing instead of pulling the oar, ' into a mooring place, they threw out the anchor stones 'from the bows'and made the stern fast'to the shore. Part of the reason for the up-curving stern of ancient oared ships may have been this practice,now known as the Mediterranean moor. The trieres was moored after each outing in a similar Homeric way.

....

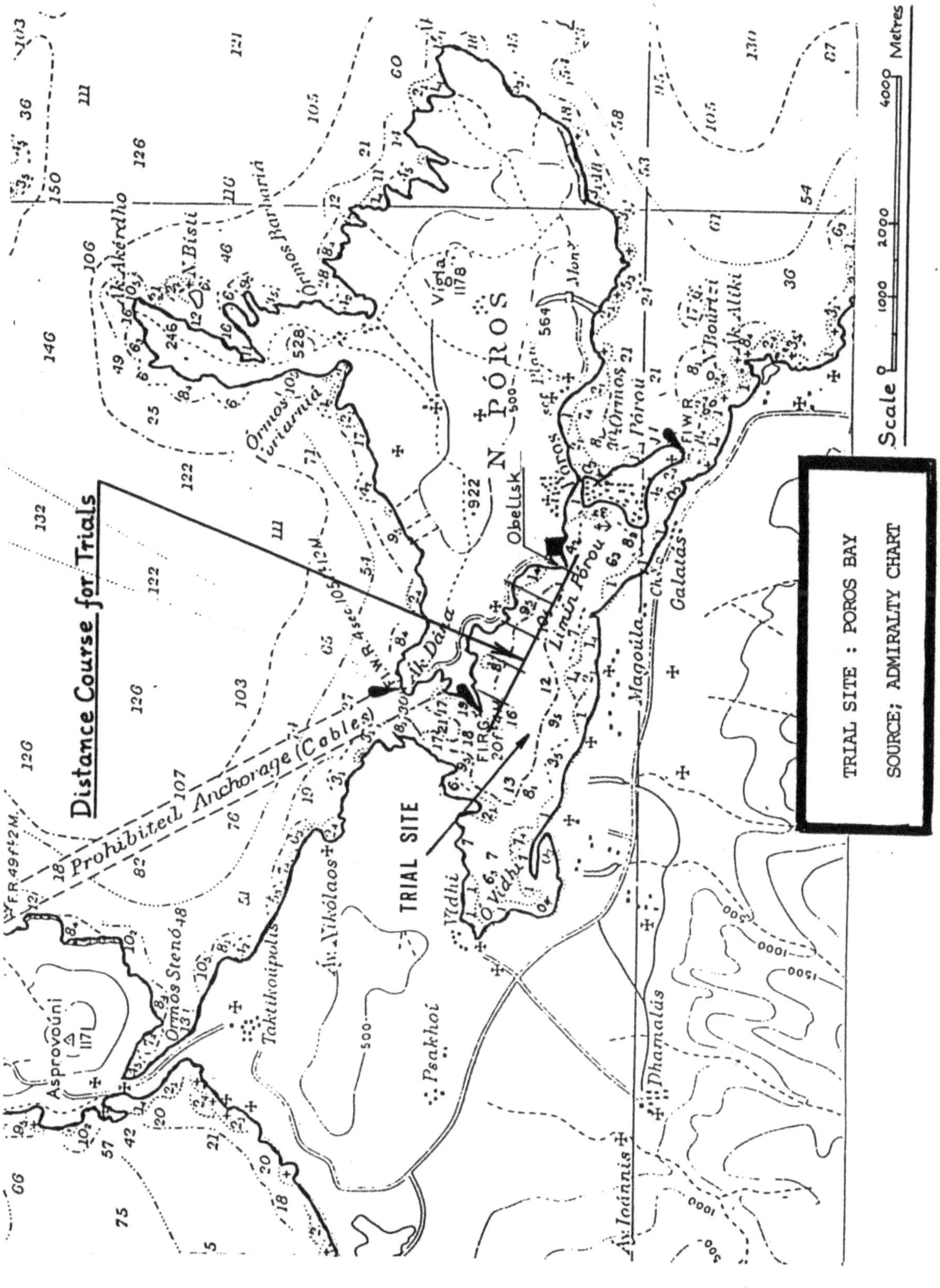

Figure 35 Chart of Poros

A projecting jetty and two buoys moored off the quay made it more difficult. But these difficulties had corresponding advantages, an inflatable rubber dingy with an outboard motor to handle the bow lines. *Olympias* always approached her mooring from the south under thranite oars only. After being rowed parallel to the quay she turned to port and a line was taken from the bow to a buoy, with the thranites at easy. Once the bow lines were secured she was backed,and stern lines (*prumnesia*) were taken ashore and secured. Backing was hard work for the thranites on their own, but by the end of the trials the Helmsman, Timothy Shaw, was able to judge the manoeuvre very well so as to keep the backing to a minimum. Earlier in the trials Captain Papadas had commanded the ship during the mooring operation but he subsequently passed control to the Helmsman. Mooring was of course less easy when the wind was blowing, but since we never experienced anything greater than a moderate breeze it never became difficult.

On outing 20, the all-day outing, *Olympias* was moored stern to a beach (Ormos Voriania, Fig.35) with two anchors from the bows to hold her off. Breast lines were also taken ashore since the anchors were considered too light, but in fact when the ship was putting to sea again the deck crew had some difficulty in breaking out the anchors. Apart from the use of breastlines, the method adopted on this occasion was the way in which a trieres on campaign would have been moored for the midday meal or for the night.

Epibatai (deck soldiers and other deck passengers)

The Athenians only carried ten hoplites (see Glossary) and four archers on the canopy-decks of their triereis rated as fast. Other triereis rated as *stratiotides, hoplitagogoi* carried forty extra soldiers.In confined waters the Athenians at the siege of Syracuse (Thuc.7.62) were driven to embark javelin men and extra hoplites and archers on their ships. Trained javelin men apparently threw their javelins sitting down, and inexperienced soldiers moving about on deck were likely to disrupt the rowing(Thuc.7.67). There were no javelin men or hoplites on board *Olympias*, but there were ten Hellenic Navy crewmen, rowing coaches, recorders, ancient historians, and perhaps the best modern equivalent of inexperienced javelin men, representatives of the world's media. It was therefore possible to make observations of the effect that different numbers of people carried on deck, *epibatai*, had on the operation of the ship.

The normal complement of the trieres was 200. The number on board *Olympias* varied from 195 to 220, with the most common figure about 200. 170 of these were rowers, and since some of the Hellenic Navy crewmen and coaches were below during outings, there were usually not more than fifteen people on the canopy deck. Nevertheless, even with these small numbers it was necessary, if the operation of the ship was not to be affected, to require deck passengers to distribute themselves evenly about the centre line of the ship to keep her properly trimmed. If a man weighing 80 kg. moved, say transversely, to the edge of the deck there was a perceptible alteration in heel (calculated to be 0.3°). People moving about the deck and talking loudly were enough to interfere with the communication of commands and disturbed the rowers' concentration. Least trouble was caused if deck passengers remained seated and silent while the ship was being manoeuvred. Standing passengers interfered with the Helmsman's line of sight. At no time were the passengers cramped for room but inevitably as their numbers increased it became more difficult to prevent them obstructing the ship's operation.

A chart of the waters around Poros is given in Fig.35.

CHAPTER FOUR : THE SCIENTIFIC AIMS

THE SCIENTIFIC AIMS OF THE TRIREME PROJECT : Boris Rankov

4.1. OBJECTIVES

The four aims of the Trireme Project have been set out on page 8 of *AT* as follows:-

1. to prove that the reconstruction designed to the historical requirements and built in accordance with the archaeological evidence has a performance consistent with historical accounts;
2. to improve understanding of naval operations in the Mediterranean from the fifth to the third centuries BC;
3. to broaden appreciation of the technical, economic and naval achievements of Hellenic society and culture from the fifth to the third centuries BC by exhibiting the reconstruction to the public in Greece with explanatory material. This purpose will, it is hoped, be amplified by housing the ship in a reconstruction of a Piraeus ship-shed of the fourth century BC from which it will be launched for historical research and demonstration at sea;
4. to recreate one of the major artefacts of Hellenic civilisation and a unique ship type of outstanding interest to naval historians and architects.

4.2. ACHIEVEMENTS

Of these aims 4. has been achieved, at least in part, by the construction and launching of the ship, and 3. is being and will continue to be achieved whenever the ship appears in public. 2. is a direct corollary of 1. and both depends upon and to a certain extent determines the conduct of the actual trial of the ship.

The results of the 1987 trials have been presented in detail in Chapter Three. What was achieved towards satisfying the project's scientific aims, that is to say, what they have contributed towards answering historical questions about the ancient world, can be briefly summarised here, and an indication can be given of what further work is needed in this field. First, and most important, is the demonstration that the three-level oar-system could and did work in practice. Despite the difficulties occasioned by heavy oars, breakages and ill-assorted and inexperienced rowers, and despite the warnings of ancient writers that rowing a trieres was no casual occupation (e.g., Thuc.1,142.6-9), reasonable progress was being made in a few days and a speed of 7 knots achieved within a week. Given that the supposed impracticability of a three-level system has always been the principal argument against it (e.g.,W.W.Tarn, *JHS* xxv (1906), 137-73, 204-24, and ***MM*** xix (1933) 52-74), this is and will remain, the most significant result of the project.

4.3. FAILURES

The ship under oar however failed to come anywhere near to the speed expected from the ancient evidence. This was, at least partly, for reasons already given. The speeds of approximately 4 knots produced at a sustainable light pressure, and of 7 knots in a single short burst, clearly fall well short of the 7 knot average which the ancient sources suggest was possible (see p.11 above). The ship proved impressive in acceleration and manoeuvrability but the factors which restricted the speed certainly affected these aspects of performance also, so that optimal figures may be expected to be rather better than those already achieved. Turning ability may furthermore be improved if a crew can be trained to a level of skill and confidence to be able to hold the ship on one side during turns. There is room also for improvememnt in the competence of the crew to back the ship. Other aspects of the ship's operation were also insufficiently investigated, in particular embarkation, where only one brow was available, and the effect of heavily armed footsoldiers moving around the ship. The brief amount of time devoted to operating the ship under sail proved remarkably successful, with the ship achieving in light winds a speed approaching that of the maximum made under oar. The 'Mediterranean moor' was also successfully accomplished.

It is clear that apart from proving the viability of the oar-system the 1987 trials only laid the foundation for the provision of the information necessary if serious progress is to be made with the second aim. This is not entirely unexpected: the real value of these first trials lay in the gaining of essential experience of the ship, the working crew and the method of operating her, which can lead to the necessary improvement.

4.4. THE FUTURE

4.4.1. The rowing performance trials, attempted in 1987, should be repeated in order to produce optimal figures for maximum speed under oar at various pressures and ratings, for acceleration, for turning and for backing. A mechanical log should be used in addition to the methods employed in 1987, in order to provide a continuous read-out.

4.4.2. Further optimal performance figures should be obtained for the ship when travelling long distances under oar, i.e., with the crew rowing for several hours at a stretch. This can best be achieved by undertaking a long voyage over several days, with the crew going ashore and resting at midday and for the night as trieres crews did in antiquity (see *AT* Ch.6). The practical difficulties in such a programme are however very considerable.

4.4.3. Figures comparable to those in 4.4.1 and 2 should be obtained for rowing with any single level of oars and with any two levels. In particular, it is desirable to discover the effect of rowing the ship without the thalamians, whose length of stroke was found in 1987 to determine the maximum practical length of stroke throughout the three levels, and was consequently felt to be a limiting factor to the overall speed of the ship as reconstructed. Conversely, rowing without the thranites would indicate their effect on the speed of the ship and might thereby help to confirm the theory that the trieres developed out of the two-level ship by the addition of the thranites rowing through an outrigger (see *AT* 25-45,163-8).

4.4.4. In the course of carrying out 4.4.1, 2 and 3, figures should be obtained to show the effect of fatigue on the ship's performance under oar both over long distances and periods of time, and over short distances. The latter is particularly important for assessing the performance of the ship in battle and would require a special test in which the crew would be asked to row a maximum speed for as long as possible.

4.4.5. Figures should be obtained to show the time elapsed and the distance travelled whilst holding and stopping the ship by oar from maximum speed (cf.Thuc,2.91.4). This is a manoeuvre which was felt to be too dangerous to attempt with the 1987 crew, but might be practicable with a crew of higher overall quality.

4.4.6. The sailing trials attempted in 1987 should be repeated and extended to find the highest achievable speed, the greatest angle of heel consistent with safety and the best methods of handling the ship in winds of varying strength and from various quarters. In particular, the purpose and working of the foresail relative to the overall rig and in varying wind conditions are in need of investigation. A long voyage should be attempted under sail in addition to briefer trials.

4.4.7. A command and communication system should be developed consistent with the historical evidence. This involves control of the ship by the Helmsman (*kubernetes*) in the stern, through a Rowing Master (*keleustes*) and a pipeman (*auletes*) in the middle of the ship and a bow officer (*prorates*) forward (cf. *AT Fig.11 p.13*). Experience in 1987 suggests that it will be necessary, in order to develop such a system in the time available, to make initial use of team leaders for sections within the ship and of an electronic address system throughout the ship, and then gradually to eliminate these aids in favour of a system historically accredited. Such a system can be expected to make use of verbal commands (cf. Xenophon *Household Management* 21.3), gesture (cf.*AT* Fig.31) and the pipe, or an equivalent. It was established in 1987 that high-pitched sound carried best within the hull of the ship. The crew must also recognise that silence in the ship, as is the practice in river rowing, must be observed. This was not the case at Poros in 1987.

4.4.8. In 1987 the thranites, who could see the water, each coached the zygian and thalamian below them, who could not. The thranites and the crew as a whole must be shown to be able to row the ship without seeing the water since this would have been denied them in battle when sidescreens were used as a protection against missiles (see *GOS* p.302)

4.4.9 The effect of deck passengers on the trim and operation of the ship must be investigated, in particular, the movement about the canopy-deck and, on occasion, the concentration in the bows of ten fully equipped hoplite soldiers (cf.*AT* pp.109-11, 130-2). In addition it would be useful to discover by experiment why it was necessary for javelin men to throw their javelins from a seated position (Thuc.7.67.2).

4.4.10 The feasibility of various operations which the crew of a trieres might on occasion have had to perform in a hurry should be investigated, as well as the time taken over them. These include the retraction of oars in anticipation of ramming or being rammed (cf. Aeschylus *Persians* 416, Diodorus 13.99.3, Polybius 16.4.14).

4.4.11 Investigation should also be made of the effect of sea and wind conditions, not severe enough to prevent the ship going to sea, on all the aspects of her performance and operation considered in the foregoing paragraphs.

4.4.12 The achievement of the objectives set out above,together with the incidental discovery of information in the course of the trials which answers questions not formulated here, will, it is hoped, satisfy the first aim of the Trireme Project proposed in *AT*, or at least indicate what modifications of ship, crew and procedures will be required to do so. What has been done so far is the beginning rather than the end of the experimemnt. The lessons learned and the performance figures acquired, probably over several years, and possibly with more than one ship, should then provide the raw data for future historians who seek to understand the literary and archaeological evidence for the construction and operation of the trieres.

4.5. THE FINAL STAGE

4.5.1. There will then remain a final stage of the project which seeks to address, in the light of what has been learnt from *Olympias*, some of the outstanding problems relating to the 'understanding of naval questions in the Mediterranean from the fifth to the third centuries BC', the second of the aims proposed in *AT*.

As a preliminary to this final stage it will be necessary to answer some important questions, which raise practical difficulties, by making use of the reconstruction. The questions all relate to what happens to a ship during ramming, in particular (i) what happens to the oars (and oarsmen) when ships ram or brush past each other;(ii) the effect of ramming on the hulls of the two ships involved; and (iii) the effect of the collision on the two crews involved. The answers to these questions will necessarily depend on a factor which is not a matter of experiment, the angle of impact normally chosen by the ramming ship as the result of tactical manoeuvre.

Since it is unlikely that the ship will be made available for such trials, at least in the forseeable future, and since they might in any case prove dangerous to the crew, alternative methods of investigation must be devised. The first question, of the effect on the oars, can best be tackled by the use of scale models which can simulate the movement and retraction of oars, as well as imaginative observation of the reconstruction.

4.5.2. The second question, of the effect on the hulls, is perhaps more difficult. John Coates has already calculated that the hull of the reconstruction could withstand the forces involved in ramming another ship, but that if the ram became lodged in the hull of an opponent travelling at an appreciable speed at right angles to the ramming vessel, it could be twisted off (*AT* p.221-2, cf.Hdt.1.166.2). A sensible commander would clearly avoid, if possible, this angle of impact. The precise effect on the hull of the ship rammed is harder to gauge and will vary according to whether she is rammed in the bows, amidships, or astern, and at what speed and at what angle relative to the ramming ship. The Athlit ram, which provided an indirect model for the ram of the reconstruction, was evidently designed not to penetrate the hull of an opponent too far but rather to spring its planking (J.R.Steffy ***MM*** pp.229-47, esp.240-1.). The forces produced and the resultant effects of various types of impact on the rammed hull need to be calculated and, if possible, tested in trial hull-sections. In particular, it would be important to know what differential in speed between the ramming ship and the ship rammed in the stern (apparently the most common forms of ramming to judge by the apparent aims of the *diekplous* e.g. Polybius 16.4.14). If the purpose of ramming was merely to spring the planks of the opposing vessel and so swamp her, the attack most convenient to the attacker, in respect both of the effect on her of the impact and of easy withdrawal, would be to ram an opponent in the quarter when he was moving at slightly less speed in the same direction.

4.5.3. The third question, of the effect of ramming on the crews, depends again on the kind of impact envisaged. If it is the successful result of deliberate manoeuvre as described above, the effect on the crew would be minimal. But in the heat of battle many other, unintended, types of impact must clearly have occurred. The effect of these can be partly envisaged from experience of similar collisions in river rowing, although these are on a very much smaller scale. This experience suggests that the oarsmen will remain seated upright and suffer no ill-effects so long as their feet are firmly secured by clogs or straps. Even with such restraints, however, the thalamians would have to brace themselves to avoid dashing their heads against the *zyga* behind them, and the undoubted chaos which would be caused to a crew without clogs in a ramming impact head-on or on the bow quarter is the best argument for their use in the reconstruction. These impressions could be confirmed by testing with dummies, and perhaps with volunteers, on the same sort of equipment used for testing the effects of automobile collisions. Deck passengers would have had to sit and brace themselves during ramming (cf. perhaps, the seated javelinmen of Thuc.7.67.2, and the drowning of Callicratidas at Arginusae in Xenophon *Hellenica* 1.6.33.). Similar tests of their risk could be tested in the same way.

4.5.4. Once the answers to these questions have been added to those provided by the sea trials, we should be much better armed to tackle broader questions about trieres operations. It may be advisable for these to be restricted, at least initially, to attempting to understand tactical aims in the fifth to third centuries rather than aiming to reconstruct particular engagements. The latter course is beset with difficulties, some of which are highlighted by a recent, as yet unpublished, attempt by Roger Leitch of the Royal Military College of Science at Shrivenham to reconstruct part of the engagement at Naupactus in 430 BC, in which an Athenian ship being pursued by a Leucadian ship swung round a merchant ship at anchor and rammed and sank her pursuer. Although only two ships are actively involved in this incident, Leitch makes five possible interpretations of the short passage in Thucydides in which it is described (2.91). What even the limited results of the 1987 trials have allowed him to do is to show that practical considerations make one interpretation much more likely than the other four. This is undoubtedly the correst approach to the use of the data provided by the project. Our understanding of the incident and of what is and what is not a possible interpretation has been greatly increased, although in the end certainty is impossible. But this was a very simple incident, relatively clearly narrated. No such account exists of the details of any full-scale engagement, and such accounts as there are imply that trieres battles tend to begin as fleet actions and end in a multitude of single combats between ships.

Such outcomes were inevitable in view of the nature of the trieres. She was a warship armed with a single main weapon which could be used against only one enemy at a time. Her task in battle, after a carefully planned fleet manoeuvre, was to pick out and attack a single, similarly armed, adversary and strike home at the most convenient angle before her opponent could do the same to her. A successful strike made her, until she could extricate herself, more vulnerable (cf.Hdt.8.90.2.). It would then have been folly for a single ship to attempt to engage two or more opponents (e.g.engage in a *diekplous* on her own).

Nevertheless, in addition to training the individual crews, trieres fleet commanders assiduously practised fleet manoeuvres (e.g.,Herodotus6.12.1; Xenophon*Hellenica*6.2.28), and it is clear that they moved into battle with their fleets deployed to gain the most advantageous position from which individual trierarchs and helmsmen could close for single combat. The most commonly mentioned manoeuvres employed to obtain such a position were the *diekplous* and the *periplous*, the two tactics spoken of by Thucydides as those on which the Athenians most relied (7.36.4), and what may be a third tactic, the *perikuklosis*, employed by the Persian commander at Artemision

(Hdt.8.10-11),by Phormio in the Gulf of Corinth (Thuc.2.83.4) and by the Athenians off Corcyra (Thuc.3.76-8), to which there appears to have been the standard defensive response of forming a circle of ships with bows outwards. The interpretation of the *diekplous* and the *periplous* is disputed (see most recently J.F.Lazenby *Greece and Rome* 2nd series xxxiv (1987) 169-77 and I.Whitehead *ibid.*178-85) and even the aims of the *perikuklosis* and the defensive circle have still to be fully elucidated.

These are the manoeuvres to which it would be most profitable to apply the combined information derived from the literary sources and the sea trials. They represent the ideal aims of trieres commanders, and as ideals they should be much easier to understand than the details of what happened in any particular battle.

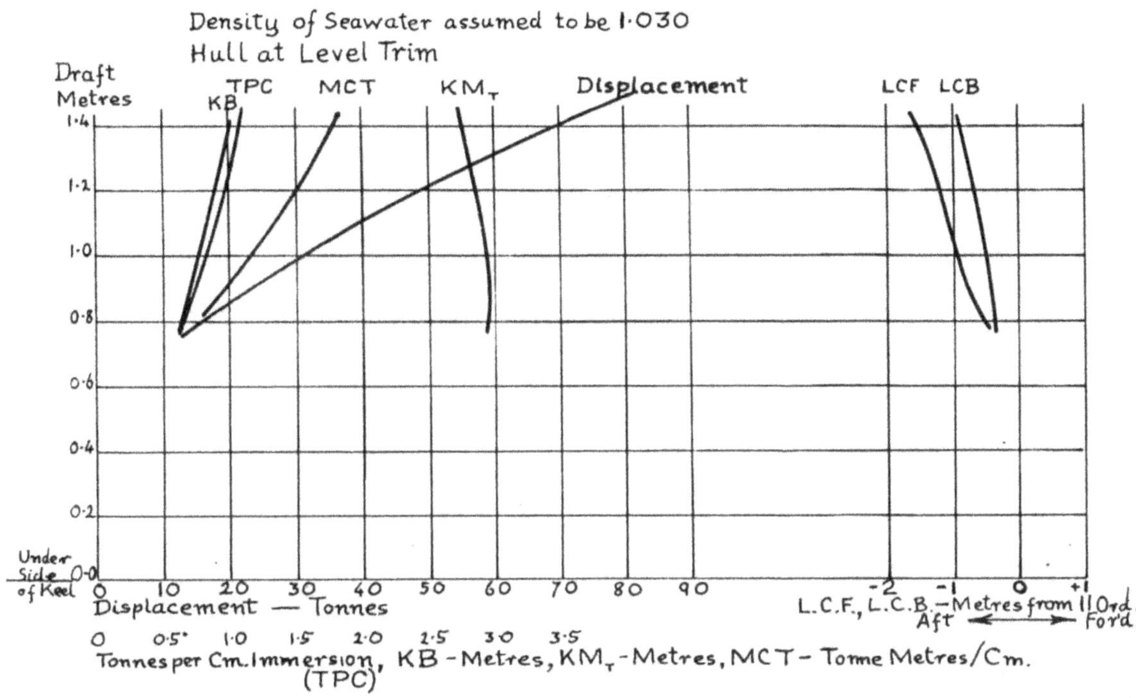

Figure 36 Hydrostatic Curves

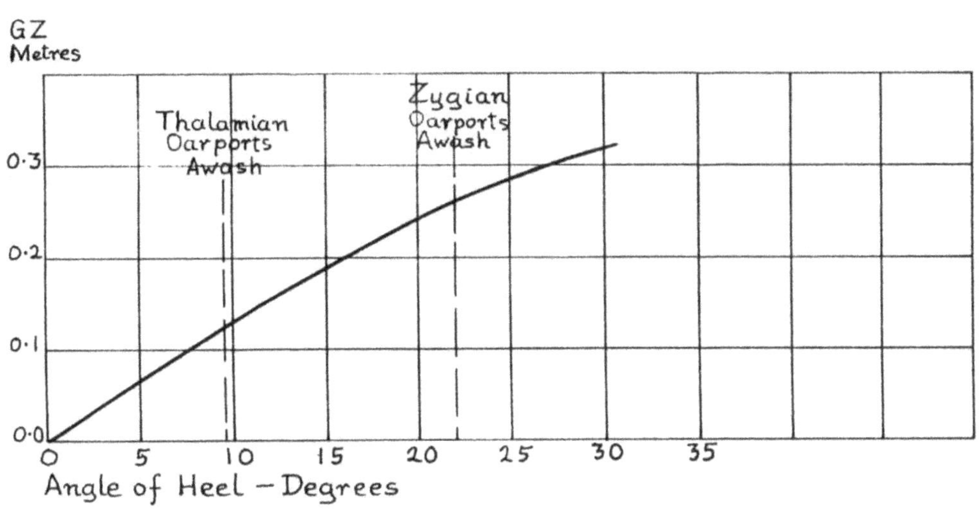

Figure 37 GZ Curve

ANNEXES : A THE SHIP

A.1 HYDROSTATICS, GZ AND PRINCIPAL WEIGHTS : J.F.Coates

1.1. HYDROSTATICS

As the displacement of the hull is light for its size, and its centre of gravity (G) relatively high, the magnitude of GM is unusually sensitive to beam (i.e. breadth of hull at waterline).

The metacentre of this hull form, which has a considerable slope outwards at the waterline (see Fig.4), is at a nearly constant height above the keel over a wide area of displacement. The hydrostatic curves are given in Fig.36 : they refer to a sea water of density 1.030.

1.2. G.Z.

This (righting lever) curve is given in Figure 36. This is an estimated curve only, being based on a calculated curve for a previous but not very different hull form (No.5) and an assumed GM of 0.75 m. It is not likely to be more than a few per cent in error for the present hull form (No.7,Mod.2) with a GM of 0.62 m., the trial condition after ballasting and with yards hoisted.

1.3. PRINCIPAL WEIGHTS

	Tonnes t.	Centre of Gravity above underside of keel m.	Vertical Moment t.m.
Hull	21.0	2.0	42
Equipment	3.8	4.2	16
Crew	15.2	2.4	36
Ballast	5.0	1.0	5
Total	45.0	2.2	9.9

A 2 HOGGING IN THE TRIERES DURING JULY AND AUGUST 1987 REPORT AND SUGGESTIONS FOR REMEDIAL ACTION: J.F.Coates

2.1. INTRODUCTION

Since leaving the building shed the hull of the trieres has hogged (i.e. the two ends have descended) by 55 mm., being the deflection upwards of the gunwale amidships relative to a straight line drawn between points near the ends of the outriggers. The deflected shape afloat has not been measured by me in detail but it seems to the eye and in photographs to follow two nearly straight lines sloping up to the mid-length of the ship, with some rounding between them at the apex and at each end fore and aft. The bow and stern remained horizontal.

The deflection of 55 mm. is about three times that which would be caused by bending on the crest of a L/40 wave and about six times that resulting from the still water hogging bending moment of 40tonne-m. This consideration and the shape of the deflection strongly indicate that the main cause is shearing slip between hull planks, the most likely cause of which is slippage of tenons in their mortices owing to a lack of tight fitting in the longitudinal direction. This possibility has always been a matter for concern in a long hull of this type, and the building specification is quite clear about the need for such a tight fit (see Fig.3 and 5 *Specification, English Edition*, paragraphs 303,335-6).

2.2. LIKELY SLIP BETWEEN HULL PLANKS IN 1987

The hull planking has probably sheared by 55 mm. in about one third of the length between thranite tholepins numbers 5 and 20,that is, in 6.2 m.. The shear angle would be 0.009 radians, which is equivalent to one plank sliding on the next one by 1.3 mm., the planks being about 150 mm. wide. A slide of that magnitude would be about as much as the sealant in the underwater seams could take without tearing. It also suggests that the previous injection of glue around the tenons below the lower wale during building, when lack of the required light fit was first noticed, had not been entirely successful in preventing slide between planks.

2.3. MEASURING THE SLIDE

The slide of one plank upon the next one can actually be measured by referring to vertical lines drawn on the planking and across seams above the lower wale during building and still there as far as I know. Many of these lines will be found near outrigger brackets. Horizontal steps in these lines at the seams will reveal the slide between planks. If the slide proves to be of the order of 1 mm. or more over any appreciable lengths of hull, the tenons in that area at least must be considered to have slid collectively.

2.4. HISTORY OF OBSERVATIONS AND ACTIONS ON PLANK FASTENINGS DURING BUILDING

In December 1985 I saw that fastenings in the keel would not be tight in the fore-and-aft direction and brought the fact to the attention of the builder and overseer. As the keel is remote from the flexural neutral axis (see Glossary) of the hull and is in a place where shear stresses would be small, I took no further action. In April 1986, however, lack of tightness longitudinally was apparent in seams just below the lower wale. This looseness was a much more serious matter as these seams were near the neutral axis, so I reported it to Commander Platis with a letter to the General Staff. It was subsequently decided to require the builder to fill the gaps between tenons and mortices by injecting glue. It was hoped thereby to obtain the equivalent of a tight fit in all plank areas below the waterline.

I understand that the glue used for this purpose was Cascamite mixed with water to a thin enough consistency to enable it to be injected through a hollow steel probe placed into the holes drilled for the locking pegs after knocking out each peg in turn. It had been found that Resorcinol glue was too viscous to inject. The gap-filling properties of Cascamite, particularly when mixed with water to make it thin, are not, I believe,very good, and it is possible that this injection may not have been very effective. I did not myself see the operation carried out. I understand that the builder was again instructed to fit the tenons tightly in the longitudinal direction in their mortices in all the seams remaining to be joined by these fastenings.

In my next visit (September 1986) I noticed mortices near the middle wale which were generally too long in the fore-and-aft direction (see Appendix 1). When in the following January oarports were cut, loose- fitting tenons were quite apparent. Commander Platis again discussed the matter with me on 21 January and we decided that, in view of the slow progress with the building, the need to complete in May 1987 at the latest and the builder's continuing incomprehension of the mechanics of the shell method of construction, the need for further injection should depend in the first place upon how the hull behaved in the strength test (*Specification,Eng.Ed.*, paras 1105-8), and in service.

We had noticed (as the builder had demonstrated) that tenons were tight in their mortices in the transverse direction, but had agreed that the frictional force arising from such tightness would not by itself prevent sliding between planks. After two months afloat it is most probable that such frictional resistance to movement had indeed been overcome, as expected.

2.5. REMEDIAL MEASURES

The measurements given in Appendix 1, of the clearances between tenons and their mortices, suggest that a further 50 mm. of hog is quite likely to develop if nothing is done to prevent it. Two kinds of remedial measures can be considered. The first would be to fix a steel or aluminium lattice girder, or a pair of them, between keel and canopy to provide by itself the necessary longitudinal strength for the ship. The second would be to extend the process already employed below the lower wale, namely to fill the spaces between tenons and mortices with a resin,more strongly to resist sliding between planks in future.

Metal lattice girders could well be most effective in preventing hogging, but should be rejected until all other less obtrusive measures have been tried and failed. Metal girders would be grossly inauthentic. The application of the second measure alone should therefore be considered at this stage.

2.5.1.*Injection of Resin* In view of the reported difficulty of injecting Resorcinol glue through the steel probe, I suggest that a resin more suitable for injection, gap-filling and of adequate cured strength be sought and injected through 4 or 5 mm.diameter holes drilled cleanly with new bits

18 mm.deep from the inside of the plank. These holes must intersect the edges of tenon and mortice about 30 mm.from the plank-edge. The position of each tenon would have to be located accurately: this may not be easy in all cases but it should be possible to insert a thin steel blade into most seams to find the edge of the tenon. Where seams remain close even after drying out, a broad, thin steel wedge driven into the seam would be likely to open it sufficiently for this purpose.

To reduce as far as possible the number of holes to be drilled and injected, I suggest that only those gaps which will close under hogging should be treated, two per tenon. In the after part of the hull such holes must be in the top half of the after edge and in the lower half of the forward edge of the tenon. In the forward part of the hull the holes must be in the lower half of the after edge and in the top half of the forward edge of the tenon.

If the injection treatment were to be confined to the regions of hull where sliding had occurred to date, the number of tenons to be treated between gunwale and middle wale would be about 2000 on the assumption that those regions prove to be about 6 metres long, one forward and one aft. Frames and other obstacles will prevent the injection of all the tenons in a treated region, though it should be possible to inject from outside the hull many of the tenons inaccessible from within. It would be particularly important to treat as many as possible of the tenons in the seams between the thalamian oarports in view of the large interruptions to the planking at that level.

The treatment must be supervised by a professionally qualified officer to ensure that it is carried out effectively. Accuracy and method of mixing the resin constituents, temperature and pot-life before curing may be critical. As the greatest bearing pressure between tenon and mortice is developed at or near the plank-edges, it is essential that the resin reaches the seams: it must therefore be seen to do so. The spaces to be injected must be dry: drillings should be first inspected to check their dryness.

2.5.2. *Choice of Resin for Injection* It is clearly essential to use a tough, gap-filling, water-proof resin. No material whose volume changes while setting or curing will do, viz. nothing with any constituent which evaporates during settling could be acceptable.

A self-curing, two (or more)-part, gap-filling, water-proof resin is needed, formulated to combine sufficient fluidity after mixing to penetrate passages about 2 mm. wide in their smallest dimension, a working time during which it can be injected after mixing of 8 hours if possible (to obtain which a curing time of several weeks or months could be accepted), and crushing strength when cured of at least 30 N/mm^2.

2.5.3. *Extent of Treatment*

To keep the crushing stresses acting on the treated tenons below the level at which permanent compression of them, and therefore looseness of planking, could develop in service, the treatment should be extended below the lower wale. This suggestion rests on the belief that the injection treatment applied during building was, or has become, ineffective. In view of the greater obstructions to treatment below the lower wale, I suggest that tenons below that wale be treated where accessible to a total number equal to the number treated above the lower wale.

2.5.4. *Bending the Hull during Treatment*

If the hull were caused to sag, i.e. the bow and stern were raised relative to the middle, while being treated, the following advantages would be obtained:

1. The gaps to be injected would be widened, easing the flow of resin to the seams and to the end of the tenon.

2. Many tenons would be caused to press against the sides of their mortices opposite to those being injected, so that after treatment they would be firm in both directions, adding to the general stiffness of the hull.

3. If the hull-shape is initially sagged, any hogging that may subsequently develop will be less obvious.

I therefore suggest that if the treatment here discussed were to be carried out, the ship should be put on blocks adjusted in height to cause the hull to sag under its own weight. I recommend that the hull be made to sag a little more than it is hogging at present when afloat. To do that, blocks would have to be set higher than the heights necessary to accord with the shape of the keel in the *Table of Offsets* for the hull (Schedule 2 and Annex 12 of the *Specifications* and in the *Lines of the Hull*, Plan 7).

The hull should be kept, after treatment, undisturbed by any movement until the injected resin has cured completely. This should be checked by testing a sample of resin used for the last injection performed, before the ship is moved for the first time after treatment. The sample must be kept during the whole intervening period in the ship to ensure that it has had the same temperature history. If the sample shows that it has not cured (by a test approved by the manufacturer) the ship should not be moved until it has passed this test.

In my opinion it would be best if during the treatment and while the resin is curing, the *hypozoma* were kept at a tension of 12 tonnes.

The extent to which the hull can safely be sagged is limited mainly by the capacity of the polysulphide sealant in the underwater seams to accommodate shear deformation resulting from plank slide. Underwater seams should be inspected for damage to the sealant when the hull has sagged 60 mm.

J.F.Coates 25 September 1987

Note added November 1988

[The Hellenic Navy carried out these suggestions during the winter of 1987/8. During the trials held subsequently at Poros in 1988 no hog was observable on the starboard side, and very little on the port side of the ship. This indicates that with properly fitted plank fastenings the trieres hull will hold its shape longitudinally. The injection, though unauthentic, has done no more than achieve a tight fit between mortices and tenons in the longitudinal direction, which by specification was to have been achieved by crushing at the specified interference-fit between mortices and tenons in this direction. This action has not affected the achievement of the Trust's aims in relation to the trials.]

Appendix I

Measurements made of tenons and mortices during the building of the trieres :25/4/86

Mortices observed to be 40 mm.wide (in the longitudinal,fore-and-aft direction) generally but varying from a few at 44 mm. to a few at 38 mm.wide. I measured tenons to be between 35 to 37 mm.wide, the specified width being 37.5 mm.

13/9/86.

In the top edges of the planks near and around the level of the middle wale I counted : -

13	mortices	36 mm.	wide in the longitudinal direction.	
50	"	37 mm.	"	"	"
26	"	38 mm.	"	"	"
1	"	39 mm.	"	"	"
1	"	40 mm.	"	"	"

As tenons were 35 to 37 mm. wide then a gap of 1 mm. between 1 and 1.5 mm. could appear under a modest shear force applied to the hull, before a majority of tenons would begin to bear on the ends of their mortices. A further slide of about 1 mm. must be expected to be necessary to cause most tenons to be crushed hard against the end surfaces of their mortices and thus resist further plank-slide more stiffly.

A 3 REPAIRS TO THE TRIERES DURING AUGUST 1987 : Paul Lipke

3.1. Initially, the number of repairs to rowing furniture needed was high, as many as 79 individual items after the first two outings. By the end of the fortnight this number was reduced to about six items at each outing, most being fairly minor. For the record, what follows is a typical daily repair list from the middle of the second week of the trials: two loose tholepins, two stretched strops, stretcher rail loose, outrigger pulling loose from some hull brackets.

3.2. The most significant causes for the high volume of repairs were :-

(1) too few fastenings and of the wrong size, improperly applied, particularly in the oarsmen's furniture. Under the extra forces unavoidable when rowers and coaches are learning as they go, these soft-drawn copper rivets, used as nails in foot clogs, stretcher rails and seats, inevitably pulled loose. As the six man repair crew located, then replaced or augmented, these fastenings, the number of failures dropped accordingly.

(2) Tholepin carlings were not properly bored or bevelled, resulting in loose tholepins, chafed oarlooms and loose carlings. Even where the initial fit of the pin in its hole was somewhat tight, the leather strops effectively corkscrewed these out of the carlings. As a stop-gap repair, 85% of the thalamian and zygian tholepins were progressively removed, their buried portions were built up with masking tape, and the pin was then driven back into its socket with mauls and wedges. In some cases inadequate bevelling of the carlings prevented thalamian oars from sliding outboard fully to where the loom shoulder should rest against the tholepin. For many thalamians (and a few zygians) this meant pulling on an oar whose handle was further inboard than was comfortable. It also caused some difficulty in keeping the blade pitched properly.

(3) The leather strops stretched more than was reasonable and acceptable owing to their being of only half the specified thickness. They were replaced progressively using rope grommets manufactured in the barracks by volunteers during afternoon rest periods. Since during the stroke the loom pivoted against the strops (and not the tholepin) good strop fit was essential to efficient rowing. It must be candidly admitted that the strop and tholepin problems helped to fire some considerable (if inconclusive) debate as to whether we should (contrary to Mediterranean practice) be rowing against the pin. [There is some evidence from the early Mediterranean that rowing was done there on occasion against the thole: see *IJNA* 7,1.(1978) p.78 fig.2 and 8.2 (1979)156-166.Ed.]

(4) Only six oars were broken during the fortnight. If a zygian blade was not shipped completely (i.e. was not fetched up fully against the oarport and tholepin) the thranite blade sweeping past would catch the end of the zygian blade and snap it cleanly off where it joined the shaft. This happened three times. Other breakages consisted of split blades or blades whose edges broke off as a result of clashes in the first four outings.

3.3. Since most of the repair problems, except for broken oars, were the result of deviations in execution of the design, their details are of little use in evaluating either the Trust's design or the challenges facing ancient shipwrights. They do however show reason why a shipwright (*naupegos*) was regularly carried on board Athenian triereis. To withstand the wear and tear of rigorous campaigning the workmanship in ancient triereis had to be of high standard, but at no time were the forces applied to *Olympias* such as to demand boatbuilding skills beyond the expected resources of an ancient naval dockyard.

9.4. Having said this I must add that in ten years in traditional boatbuilding I have never worked on a less user-friendly craft. Climbing over, around and under shipped oars, seats and stretchers left the repair teams frustrated and contorted. Movement within the moored ship was so slow that it was helpful to assign two people to carry every repair item, one to climb into position to do the work, the other to assist and run for the needed materials and tools. The repair crew consisted of : Paul Lipke (in charge), David Wigg, Richard Delacey, Andrew Miller, Ford Weiskittel, Charles King, Phil Metcalfe, Jeff Moggeridge, Steve King, Andrew Hill and Jamie Elwin.

4.1. LIST OF PARTICIPANTS: POROS 1987

Trierarch: Lt.Dimitrios Papadas Hellenic Navy
Deck officers and deck crew:Hellenic Navy
Rowing Master: Dr Michael Budd
Assistant Rowing Masters:Dr S.J.Walter, T.Watson, P.MacLeod
Sailing Master: O.T.P.Roberts
Helmsman: J.T.Shaw
Chief Recorder: Ian Whitehead
Assistant Recorders: P.Lipke, M.Lipke
Photographer: J.S.Illsley
Interpreter: Y.Manuelides
Repairman: P.Lipke
Barrackmaster: P.MacLeod
Assistant Barrackmasters: A.D.I.Green, M.Lipke
Team Leaders: M.G.Wilmot, B.Johnson, C.T.Cox, A.P.Ruddle, M.Jolly, P.Thompson.
OARCREW: John Aiglewood, Henry Fraser Alexander, Jeremy N.M.Allen, James Alston, Matthew Amies, Peter N.D.Askew, Neville Ian Aston, Kevin Philip Aston, Geoffrey Austin, Timothy J.Austin

Nicholas Badman, Simon Badman, Amy Baird, Helen Baird, Lucy Baird, Fiona Baker, James Bamforth, T.J.Beaver, Eeva Berglund, Sarah Binns, Alison J.Blunt, Toby H.J.Blake, Sharon Bowley, Peter M.Brett, Stephen Brindle, Neil Brisley, R.Ivan Brooker, John A.Brooks, Sylvia F.Brooks, Carolyn Brown, Keith Brown, Paul I.Brown-Kenyon, Erica C.Bruce, William J.M.Bruce, Peter Burdess, Adrian G.Burns, Douglas Busvine.

Andrew M.Callard, Richard Charrington, Geoffrey S.Chatas, Paul.R.Clarke, Jonathan Cloke, Vanessa J.Collingridge, Diana J.Coop, Julian Cooper, Matthew Colvill, Erica Cox, Richard M.Crooks, Ian E.Crompton, Tim J.A.Crow, Robert M.Czwarno.

Robert Dangerfield, Michael P.Darasz, Catherine Davis, Michael P.Deards, Richard de Lacy, Guy N.Deslandes, Andrew P.Dickerson, Sally Duckett.

Bruce W.O.Eadie, James Elwin, Fiona E.J.Esson, Hugh W.Evans.

Brian M.Fenner, Graeme Fife, S.R.Fitch, Catriona Fox.

Douglas Galbi, William Green, Jim Gregory, Eleanor Grey, Douglas J.Gurr.

Ben Hall, Paul Hallett, Philip S.Hallows, Chris Hamer, Rupert Hare, Toby Harnden, Dr Andrew Harrison, Tom Hassall, Chris W.Hassall, Edward C.Hassall, Laura Hawksworth, Stephen Higgins, Maisie Hill, Andrew C.H.Hill, Nigel Hill-Paul, Bridget Holland, Tom Hooper, Patrick A.J.Horton, Adam Howard.

Mrs Maria C.Ioreva, Alison Ironside Smith.

Adrian Johnson, Gavin Johnson, Alasdhair W.Johnston, Sunniva Jordan.

Alex Kershaw, Charles King, Steve King, Jane A.Kirby.

Simon J.Lindsay, Nicholas J.Lowcock

Jean McCollister, Rupert Macey-Dare, Archie McConnel, Conrad M.McDonnell, John B.W.McDonnell, Susan McDonnell, John MacGinnis, William D.McInnes, Dr James MacLeod, David MacLeod, I.David McNamara, Paul S.MacPherson, Kevin McWilliams, W.J.Manning, Paul Markwick, Dominic Martin, I.P.Mavromattis, Philip W.G.Metcalfe, Andrew Miller, David Miller, Alison Mills, Henry Milner, Geoffrey Moggridge, Graham L.Mornell, James Morris.

Helen Newsam, Richard Norton, Trevor Nott.

Erica Othen.

Sophie Painter, Julia Parkes, Jack Payne, Mark Payne, Huw Peach, James W.B.Pipe, Gregory S.Pipe, Christopher Polack, Dr Fiona Polack, Konrad Pollard, Richard Press, Graeme L.Proudfoot.

James W.Rankin, Philip J.Rawlins, Matthew Rhodes, Jason J.Richards, Paul W.Richards, Penelope l.Richards, Martin Richmond, Julia Riddlesdell, Matthew Ridgewell, Sean Ringstead, Chris Roberts, Chris E.Roberts, Deborah Rogerson, Jim Ronaldson, Robert Ronaldson

Michael Scholes, Ian J.Sharp, Sean J.Sinclair, John M.Snell, Stephen Spowart, Jo Spreadbury, Henry J.A.Stanford, Mrs Joan Stanley-Baker, Michael Stanley-Baker, Richard Stanley-Baker, Colin Stringfellow, Brigid A.Sutcliffe.

Sarah Talbot, Bruce Thompson, Simon Tiplady, Theodor I.Troev, Philip G.Tully, Mrs Caroline Turnbull, J.A.Turnbull.

Tiffany Vignoles.

Michael A.Ward, R.Clare F.Wardle, A.Shan Wareing, Vincent Warner, Peter Watson, Allan M.Watt, Prof.S.Ford Weiskittel, Helen Whitaker, Rupert White, Sally White, Emily Whitehead, Peter Whittall, John Whitting, David Wigg, Nicola A.Wilberforce, Justin Williams, Tim Wright, Richard Woods.

Mike Zeidler.

The following members of the Trireme Trust :- John.Coates, Captain Colin MacGregor Royal Navy, Yannis Manuelides, Rosie Randolph, John Morrison, Frank Welsh.

4.2. PROGRAMME OF TRIALS CARRIED OUT

DATE	NO. TIME	PRINCIPAL EXERCISES	SIGNIFICANT RESULTS
1.8.	9.46-11.05	Embarkation,Manoeuvring Separate sections,all oars	Max.speed 4.18 kn. Embarkation 21 mins
1.8.	2 19.08-20.24	Manoeuvring all oars Backing,catching crabs	'Not very successful session' Log
2.8.	3 9.08-11.01	Spinning boat with oars only Rowing with thranites only then zyg.and thal.,then all.	Embark. 12 mins Thranites 3.75 kn.. Thal.& zyg.3.36 kn. Disembark. 12 min.
2.8.	4 18.59-20.17	Rowing in sections.Turning difficult in F.4 wind.All oars.Thr.told to find slot. betw.zyg. and thal.oars.	Headway into F.4 All oars into wind 3.89 kn..Downwind no sails 2.28 kn.
3.8.	5 8.53-10.38	Whistle for emergency easy. Sheepskins issued.Thal.in sects.then together to lengthen stroke.Zyg and thal. in sects.Then together.3 lev. in sects.Then together.	3/4 press.32 spm 5.48 kn.Light pr. 27-9 spm 4.98 kn. All together 20s at 3/4 pressure.
3.8.	6 18.43-20.16	For max.body swing thal. to bring heads up to crossbeam. Bt.sail 5mins.Rowing in comb- inations of sep.levs.& sects. All oars 5min.lt.pr. some fm pr.	Lt.pr.28spm 5.48kn Fm.pr.30spm 5.71kn
4.8.	7 8.55-10.22	Rowing in combs.of sep.levs. All oars.Turning.Lt.& fm.pr. Fm.pr.test.Acceleration trial.	All lt.25spm 5.16kn All fm.29.6 6.30kn Thr.fm.32-4 5.13kn Zyg.fm.34-5 4.36kn Thal.fm.31 4.01kn All.fm. 38 6.95kn
4.8.	8 18.42-20.03	Backing with zyg,and thr.in sects.then together.3 levs. sects.then together.Lt.pr. Turning at lt.pr.all oars+ rudders. More lt.pr.	Bkg.zyg,thr. 2.40kn Bkg.all oars 2.43kn Lt.pr.tn.+rdr 90dg in 50 secs.29-30 spm:180dg.1m.20secs
6.8.	9 8.59-11.00	2000m course at lt.pr. Row out of bay.2/3 crew.Back in sects.continuously.	Course:12min.55secs 2/3 cr.lt.pr.4.26kn Voyage est.10,000m.
6.8.	10 18.46-20.07	Turning trials.1.Strbd oars	1.90dg.in 58 secs

		backing,pt,oars paddling. 2.Fm.pr.inside oars going easy. 3 min.fm.pr. Few emergency stops now.	2.90 dg." 35 secs 180 dg." 61 secs
7.8.	11 9.05-10.04	Ship rowed in sects.1/3 crew then ½ cr.at one time,then 2/3 cr.Fm.pr. All oars fm.pr. for 1 min. Bow cr.fastest.	Bow 1/3 : 4.28 kn. Bow ½½: : 5.20 kn. Bow 2/3 : 5.68 kn Fm.spm now c.40-44.

FIVE TONNES OF BALLAST ADDED TO RAISE WATERLINE FORWARD

8.8.	12 9.05-10.23	Experiments with low spm, separate levels.All oars lt.pr. Turning.Acceleration,5 str.at lt.pr.,10 at3/4,15 at fm.	Spm rising to 39 in accel.runs
8.8.	13 18.44-19.51	Accel.practice.10 lt.,10 3/4,15 fm. Bkg with all oars no success. Bkg with zyg,and thr.Turns at fm.pressure.	90dg rev.turn from stop 75 s. Bkg z.thr.2.59kn
9.8.	14 8.37-10.32	5 min.run at lt.pr.Then ½ crew rowing at a time for 1hr 40min. Out of the bay.	½cr.lt.pr.c.4kn

BALLAST REDISTRIBUTED.BOW SLIGHTLY RAISED

9.8.	15 19.04-20.32	Sailing outing. Main & boat sail. V.lt.wind died.Back under oar.	Wind on strbd quarter:2.34kn
10.8.	16 8.51-9.57	Four 500m.runs at fm.pr. Accel.tr.5str.lt,10 3/4,15 fm.	
10.8.	17 18.43-20.16	Accel.tr. spm rising to 43½.6 Turns to pt.& strbd.Proper accel. run.	6 kn.in 30 secs
11.8.	18 8.56-10.03	Two 1000m.runs.1.at 3/4pr, 2.fm.pr.Two 180 dg.turns.	1.33spm 5m.43s. 2.32-4spm5m.26s
12.8.	19 19.05-20.23	Sailing.Wind F.2-3.Main & Boat s. Swift sail to end of bay.Sails brailed after 20 mins.Rowed back in sections.Short run at ½pr.	5.88 kn.with wind on qur.
13.8.	20 10.18-12.28	All day outing.Rowing in sects. Short period all oars and b.sail. Mooring at beach.Disembarkation by 2 stern ladders.Back partly under sail (wind 10-15kn),partly with all oars 10-15 min.runs	Sailing speed 6kn on broad . reach,both sails set
14.8.	21 9.00-10.15	Two 500m.runs at fm.pr.,then lt.pr. through channel past the town.Turn and back to mooring under oar.	

ABBREVIATIONS IN TRIALS PROGRAMME

accel.= acceleration	fm.= firm	sect.= section
betw.= between	kn.= knots	sep.= separate
bkg.= backing	lev.= level	s.p.m.= strokes per minute
comb.= combibation	lt.= light	str.= stroke
cr.= crew	max.= maximum	strbd= starboard
dg.= degree	min.= minimum	thal.= thalamian
embark.= embarkation	pr.= pressure	thr.,thran.= thranite
est.= estimated	pt.= port	zyg.= zygian
f.= force	qur.= quarter	

5.ROWING TABLES

Outing Number	Wind Speed (Beaufort Scale) +=tail,-=head	Rating (spm)	Speed (knots)
5.1.	TABLE 1 : ALL OARS AT LIGHT PRESSURE		
1	+3	27	4.18
1	-3	27	3.44
2	-2	..	3.04
2	+2	..	3.63
2	+2	..	3.72
5	0	27-29	4.98
6	-3	28	4.76
6	-3	28	5.07
6	-3	28	4.60
6	-3	28	5.48
6	-3	29	4.60
7	+1	25	4.97
7	+1	25	5.08
7	+1	25	5.16
8	+2	29-30	4.87
8	+2	29	4.87
9	0	30	5.09
9	0	30	5.36
9	0	30	5.12
9	0	30	5.11
9	0	31	5.08
9	0	31	5.55
ADDITION OF BALLAST AFTER OUTING ELEVEN			
12	0	28	4.08
12	0	28	4.54
13	-1	32.5	4.21
13	-1	32.5	4.86
14	0	30	4.66
14	0	31	4.88
14	0	31	5.11
14	0	32	4.86
14	0	32	4.89
14	0	32	4.91
14	0	32	4.95
17	+3	34	5.48
17	+3	34	6.09

5.2. TABLE 2 : ALL OARS AT THREE QUARTER PRESSURE

4	-4	..	3.89
5	0	32	5.48
ADDITION OF BALLAST AFTER OUTING ELEVEN			
18	0	33	5.37
18	0	33	5.26

5.3. TABLE 3 : ALL OARS AT FIRM PRESSURE

6	-3	30	5.71
6	-3	30	5.65
7	+1	29	6.07
7	+1	29	6.30
7	+1	29	5.07
7	-1	38	6.95
10	+3	34	6.46
11	0	40	6.05
11	0	40	6.51
11	0	40-44	6.19
ADDITION OF BALLAST AFTER OUTING ELEVEN			
12	0	39	6.35
13	0	38	5.97
16	0	39	6.09
16	0	39	6.09
16	0	39	6.09
16	0	39	6.02
16	0	39	6.41
16	0	39	6.09
16	0	38-43	6.30
17	-3	40.5	5.55
17	-3	40.5	5.27
21	-2	34-36	5.94
21	-2	36-38	6.00
21	-2	36-38	6.14
21	-2	38	5.86
21	-2	38	6.02
21	-2	38	6.11
21	-2	38	6.06
21	-2	38	6.38

5.4. TABLE 4 : SEPARATE LEVELS

Pressure	Wind Speed Beaufort Scale	Rating S.P.M.	Speed knots
(a) Thalamian Level			
Light	+2	20	3.10
Firm	0	31	4.01
(b) Thalamian and zygian levels			
Light	-2	..	3.30
Light	-2	..	3.36
Light	0	26	3.94
(c) Zygian level			
Light	+2	26	3.06
Firm	0	34-35	3.67
Firm	0	34-35	4.36
(d) Thranite Level			
Light	-2	..	3.75
Light	+4	..	4.21
Firm	0	32-34	5.01
Firm	0	32-34	5.13

5.5. TABLE 5 : BOW HALF AT LIGHT PRESSURE

Outing Number	Wind Speed Beaufort Scale	Rating S.P.M.	Speed knots
14	-2	31	3.41
14	-2	31	3.74
14	0	30	3.80
14	-2	31	3.96
14	0	30	4.08
14	0	32	4.07
14	0	31	4.08
14	0	30	4.11
14	0	31	4.11
14	0	31	4.20

5.6. TABLE 6 : STERN HALF AT LIGHT PRESSURE

1	-2	22	2.40
14	-2	30-31	3.18
14	-2	30-31	3.40
14	-2	30-31	3.56
14	-1	31	3.58
14	-2	30	3.67
14	-2	30-31	3.67
14	-1	31	3.75
14	-1	31	3.90
14	-1	31	3.96
14	-1	31	3.97

5.7 TABLE 7 : BACKING WATER

Outing number	Wind Speed Beaufort Scale	Rating s.p.m.	Speed in knots
(a) Zygians and Thranites			
8	-2		2.47
12	-1		2.59
(b) Zygians and Thranites facing Bow			
13	-1		2.25
13	-1		3.00
(c) Whole crew			
8	-2		2.34
8	-2	21	2.43
13	-1		1.80

ANNEXES : B ROWING THE SHIP

B 6 SPECIMEN QUESTIONNAIRE ISSUED TO ALL OARCREW AT THE END OF THE TRIALS

PLEASE BRING COMPLETED QUESTIONNAIRE TO STAFF ROOM BEFORE SUNDAY,OR MAIL TO:IAN WHITEHEAD, CLASSICS DEPARTMENT, THE UNIVERSITY, NEWCASTLE UPON TYNE ENGLAND.

NAME:SOPHIE PAINTER

ADDRESS:

OAR POSITION (S): Thalamian (zygian) Bows, Stroke side.

WHAT IS THE LONGEST CONTINUOUS ROW YOU THINK YOU COULD SUSTAIN AND AT WHAT RATING?: 3/4 hr in thalamian, rating approximately 30.

WHAT IS THE LONGEST ROW WITH RESTS YOU THINK YOU COULD SUSTAIN? : 2 hrs (at a guess!!)

WERE YOU COMFORTABLE? : Yes

HOW MIGHT YOUR POWER BE APPLIED MORE EFFICIENTLY?: A stroke long enough to take the catch with straight arms would improve efficiency.

DID THE COMMAND SYSTEM WORK ADEQUATELY? IF NOT HOW SHOULD YOU CHANGE IT? Yes but only when concentration was high ! a set number in the countdown to easy (5 or 10 <u>not</u> both) would be easier.

DO YOU HAVE ANY SUGGESTIONS AS TO HOW THE EFFICIENCY OF THE SHIP COULD BE IMPROVED? A more careful match of size and strength to position in the boat would relieve some problems with varying stroke lengths.

COMMENTS:-

B.7 ON POWER AND PACE IN THE MODERN TRIERES :
DEDUCTIONS FROM THE SEA TRIALS IN 1987 AT POROS : J.T.Shaw,February 1988.

NOTE: 1 newton = 10^5dynes = 0.10197 kgf = 0.224878 lbf
1 knot = 0.5148 ms^{-1}
A "complete stroke" is defined as consisting of a "stroke" during which the oarblades are working in the water, and a "run" during which they are being brought forward through the air.

7.1.INTRODUCTION About 120 years ago, shortly before the introduction of sliding seats, men in fixed-seat racing eights were producing power outputs that if reproduced by the 170 oarsmen in the modern trieres would propel her at a speed of some 10 knots, given calm weather. In 1987 the trieres made her first sea trials. She achieved 7 knots, an encouraging result but one that leaves considerable scope for improvement. Her acceleration trials have shown that at all the attained speeds the effective horsepower of the crew fell well short of what ought to be possible. It is thought that in future trials the effective power can be substantially increased by lightening the oars and by recruiting smaller yet still strong people who will not be so much hampered by the limitations of the space in the trieres as were the crew in 1987. The following investigation of what was achieved in 1987 is offered as a contribution to the discussion of improvements.

7.2.RESISTANCE OF THE TRIERES Trials carried out with a model of the trieres in a test tank have established how the resistance to forward motion of the full-sized ship, less rudders, to forward motion would vary with the speed. The additional resistance of rudders, partly and fully immersed, has been calculated, assuming a drag co-efficient of 0.5. Calm weather is assumed and any effect of air resistance is neglected. The results are for a trieres displacing 47 metric tonnes. The actual trieres displaced only 45 tonnes and accordingly her resistance was probably a trifle less than those determined by the model. However the latter will be used in what follows.

The resistances were quoted to the present writer in the form of curves of horse-power versus speed (Fig.37). To reduce errors of reading these curves and to permit extrapolation to low speeds the writer sought a mathematical relationship, by plotting the natural logarithm of the horsepower against that of the speed, from 4 to 8 knots. This plot turned out to be a good straight line having a slope of almost exactly 3, indicating that a cube law would make a good approximation to the facts.
A simple and sufficiently accurate law for the present purpose is this: with rudders fully immersed, the horsepower needed to propel the ship at a steady speed of v knots is 0.045 v^3 and the power in kilowatts is 0.0336 v^3 provided that v does not exceed 8. If v is expressed in ms^{-1} (not exceeding 4.1) the power in kilowatts is $0.246v^3$.

This law implies that the resistance R in newtons is given by a square law, namely:

$$R=246\ v^2 \quad (\ v \text{ in } ms^{-1})$$
$$R=65.2\ v^2 \quad (\ v \text{ in knots }\ )$$

again provided that the speed does not exceed 8 knots, 4.1 ms^{-1}. At higher speeds than 8 knots, 4.1 ms^{-1}, the power must be read from Figure 37. For example, with rudders partly raised to lessen the resistance, the power required to maintain a steady 9.5 knots, 4.89 ms^{-1}, is 26.8 kw (36 hp.); the resistance at that speed is therefore 5490 newtons (1234 lbf).

Some comparative data are available for a fixed-seat eight, that of the Oxford Etonians of 1870 who covered the Henley course in the shortest time ever achieved by a fixed-seat crew. Their time was 7 minutes 17 seconds indicating an average speed of 9.72 knots, 5.00 ms^{-1}, on the assumption that the adverse current had its normal summer value of 0.17 ms^{-1}. Boat design did not change greatly on the introduction of sliding seats, and it is likely that the resistance of the 1870 eight was close to the known resistance of eights 30 to 40 years later. If so it was in the region of 307 newtons at 5.00 ms^{-1}.

The wetted surface of the 1870 eight would have been about 9.8 m^2: that of the 1987 trieres when displacing 45 tonnes is 118 m^2, i.e.almost exactly 12 times as great. Accordingly, even though the trieres has a rougher surface than an eight, her frictional resistance per oar is probably less than that of the eight: a reason for optimism regarding her potential top speed under oar.

7.3. THE OARS OF 1987 The oars to be used near the bow and stern of the trieres were 9 cubits in length; all the rest, the majority,were 9.5 cubits. These lengths are attested by ancient inscriptions. The cubit was taken, from the dimensions of the temple on the Acropolis known as the Hecatompedon (above p.2), to be 0.444 m.. Accordingly the majority of the oars were 4.22 m.long. The shapes of the blades depended on whether the oar was to be used at the highest, middle or lowest level. The lengths inboard of the tholepins were intended to be 1.02 m. but this dimension could be and often was increased by the oarsman, as the position of the fulcrum was not very positively located on the oar. The oars were rigged forward of the tholepins, and were tied to them by oarloops of leather or rope through which the strain was taken. These oars were very strongly made and consequently were rather heavy to handle, and it has been decided to pare them down considerably in time for the trials in 1988[1]. The dimensions used in 1987 and the modifications intended for 1988 have been fully described and illustrated by J.F.Coates (see Bibliography p.97).

For the purpose of this Annex the force applied to the oar by the oarsman's hands is taken to act through a point a hand-and-a-half's breadth from the end of the handle; this implies an effective inboard lever arm of 0.82 m.in 1987. The reaction of the water is taken to act through the centre of pressure of the blade, whose position is not fixed but is taken, as a fair approximation, to be 0.30 metres from the tip. This implies an outboard lever arm of 2.90 m.in 1987. For comparison,the effective inboard and outboard lengths of the oars of the Oxford Etonians in 1870 were 0.84 m. and 2.32 m., the full dimensions being 1.04m.and 2.62m., i.e.3.66 m.overall. (The centre of pressure of the blade is again assumed to have been 0.30m. from the tip.) An oarsman rowing hard with such an oar could have moved the end of the handle a distance of 1.22m. relative to the boat per stroke in the water, and the tip of the blade would then have been moved 3.07m. relative to the boat. The mid-point between the hands would have moved 0.99m. and the centre of pressure of the blade 2.72m.. These figures are used in a later section.

7.4. MEASURED PERFORMANCE OF THE TRIERES UNDER OAR IN 1987 The Trireme Trust's Newsletter No 6 of October 1987 stated, inter alia, "The results were encouraging, as John Coates' summary shows:

Speed 1: Maximum achieved at 38 s.p.m. for a short burst: 7 kn.(13 kph)
2: For 2000 m. at 30 s.p.m. and light pressure 5 kn (9 kph)
Acceleration : From rest a speed of 6 knots was realised in a minute for most of which time the acceleration was 0.12 m/sec.^{2}".

In what follows, these sparse details are combined with the resistance data and the estimated all-up weight of the ship, to reveal the effective power output of the crew at speeds from 0 to 7 knots; and then the dimensions of the oars are brought in, to enable the total power output, the oarsmen's pull, and various other details including the law governing the slip of the oars in the water, to be deduced. Comparison is made with the Oxford Etonian eight of 1870.

7.5. EFFECTIVE POWER OUTPUT IN 1987 The effective power output of the crew is defined as equal to the power absorbed in accelerating the mass of the fully laden ship,of any entrained water, and of her crew in line ahead, and in overcoming the resistance of the water as ascertained by the tank tests of the model. Air resistance is neglected.The equation linking the resistance of the water with the speed of the ship has already been given. The acceleration is dealt with here.

It must be admitted that the data on acceleration have been obtained indirectly via some rather sparse measurements of distance and time, and that the mass involved is not precisely known, mainly through lack of information about the entrained water, though it is obvious that when the ship's speed is nil, so also is tghe entrainment. From here on the effect of entrained water is ignored. Errors are therefore likely, but nevertheless some interesting points emerge: they help to refine the data.

The statement has been quoted above, to the effect that for most of the time during a half-minute's acceleration from rest to 6 knots the acceleration was 0.12ms^{-2}, i.e., its average over each complete stroke was constant for most of the time. "Constant" must be read as meaning "approximately constant". Such a reading would be consistent with the hypothesis that the mean effective pull - i.e. the mean pull per complete stroke - exerted by the crew was constant at low speeds but fell off gradually above 3 knots and then fell rapidly above 4 knots. At speeds below 3 knots the resistance of the water was very small, so that very nearly all the effort by the crew was realised as acceleration then.

The magnitude of the crew's mean effective pull at low speeds is obtained here by assuming that at a speed of 3 knots the mean acceleration had exactly the quoted value of 0.12ms^{-2}. The force required to produce this acceleration is given by mass x acceleration, i.e.,

$$45 \times 1000 \times 0.12 = 5400 \text{ newtons}$$

The force required to overcome the resistance of the water at 3 knots is given by: $65.2v^2 = 587$ newtons

Total = 5987 newtons

In accordance with the hypothesis of a constant mean effective pull at low speeds, this pull, 5987 newtons, is assumed to have been exerted at 0,1,2 and 3 knots. On that assumption the actual mean accelerations at those speeds have been calculated. They are tabulated below:-

Speed in knots	0	1	2	3
Acceleration ms^{-2}	0.133	0.132	0.127	0.120

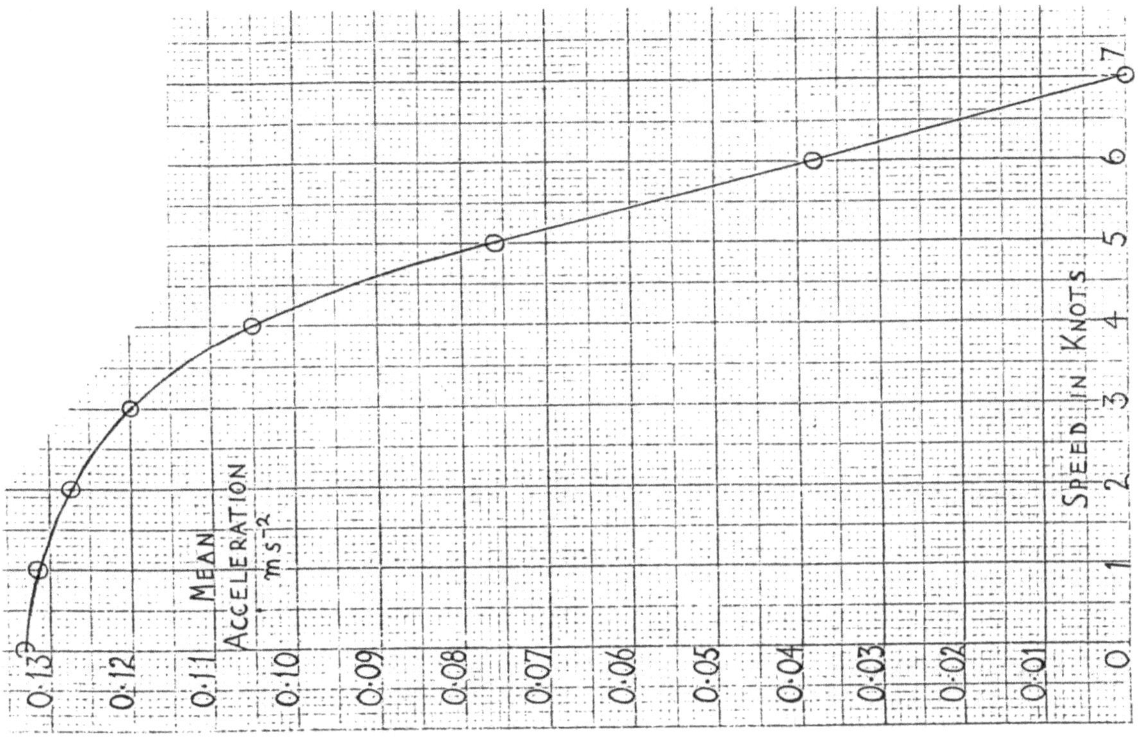

Figure 39 Acceleration - Speed

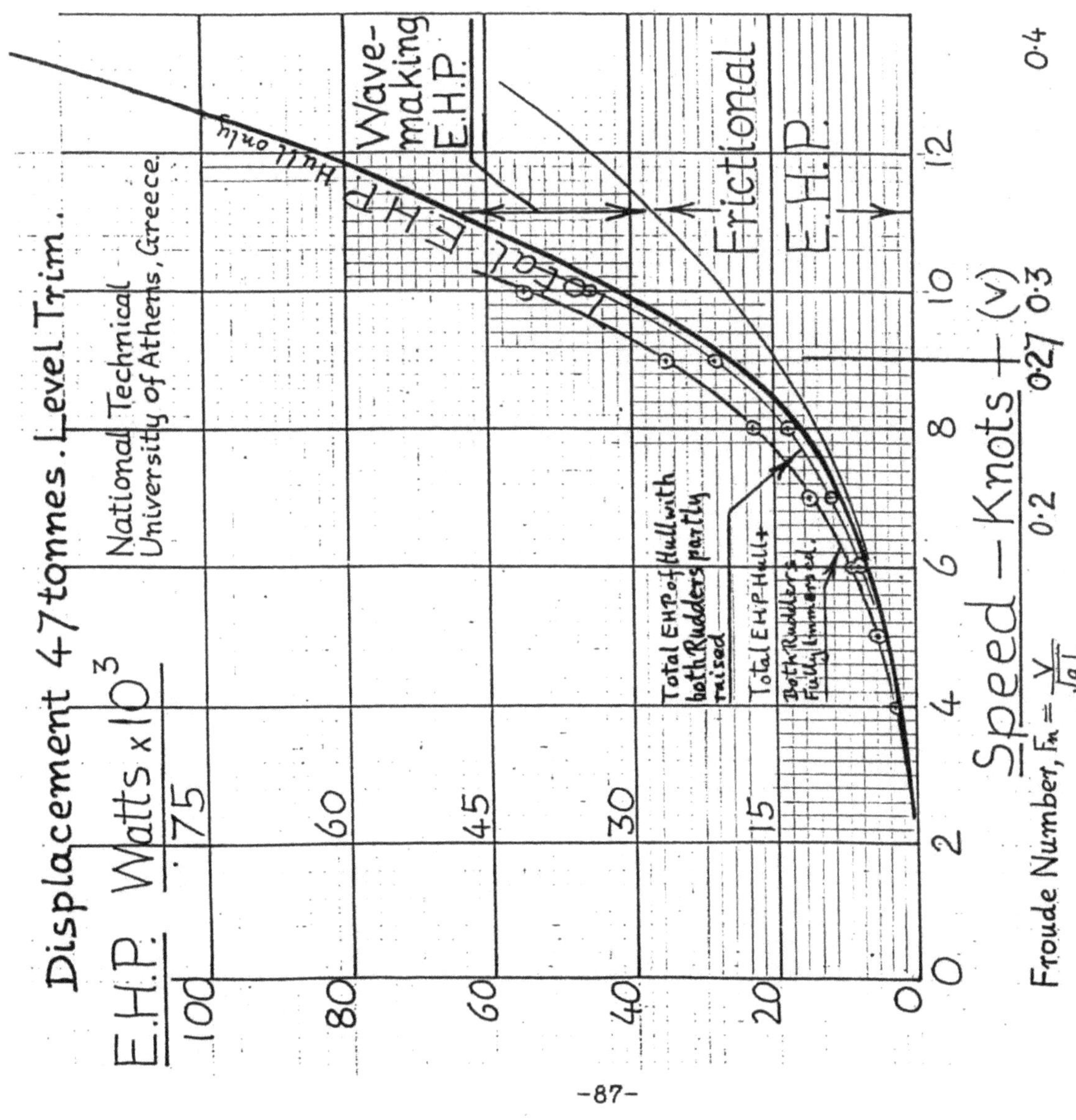

Figure 38 Effective Horsepower

On this basis a speed of 3 knots would be reached in 12 seconds from a standing start, and in a distance of 9.4m., i.e., about a quarter of the length of the trieres.

The mean acceleration is assumed to have fallen smoothly to zero as the speed increased from 3 to 7 knots. That is the basis on which the accelerations at 4.5,and 6 knots have been located on the graph, Figure 38. These accelerations agree well with the original measurements of time and distance. The diminution of the acceleration is only partly accounted for by the increase of the resistance of the water with increasing speed: increased entrainment would have had some,probably slight,effect, ignored in these calculations, and certainly the crew's effective pull diminished as the speed increased above 3 knots. This must reflect an increasing difficulty in catching hold of the water with the blades.

The mean effective pull of the crew at 4, 5, 6 and seven knots has been calculated from the resistance at these speeds coupled with the acceleration shown in Figure 38. The results are tabulated below, and shown in Figure 39:-

Speed in knots	4	5	6	7
Acceleration, ms^{-2} (Fig.38)	0.105	0.076	0.038	zero
Force producing acceleration,newtons	4725	3420	1710	zero
Force overcoming water resistance,newtons	1043	1630	2347	3195
Mean effective pull,newtons	5768	5050	4057	3195

The mean effective power output by the crew during the acceleration is given by the formula:

$$\text{Mean effective pull in newtons} \times \text{speed in } ms^{-1} \times 10^{-3} = \text{power in kilowatts}$$

The results are tabulated below:-

Speed in knots	0	1	2	3	4	5	6	7
Mean effective power, kw	0	3.1	6.2	9.2	11.9	13.0	12.5	11.5

As the trieres contained 170 rowers the greatest mean effective power per rower was about 76 watts and was developed between speeds 4.8 and 5.4 knots: see Figure 40. At the top speed of 7 knots the mean effective power per rower was just under 68 watts. In comparison with what was achieved by the best fixed-seat eights of the last century these powers are low. For example, the Oxford Etonians of 1870 covered the Henley Regatta course in the shortest time ever recorded for a fixed-seat eight. Details already given enable their mean effective power to be calculated.:

$$307 \text{ newtons} \times 5.00ms^{-1} \times 10^{-3} = 1.54 \text{ kw}$$

i.e. 192 watts per oarsman,0.257 hp

This was an average power in a race lasting more than 7 minutes. The power in a short sprint would have been higher. Other crews of the period were nearly as good. The resistance of the Oxford Etonians' boat is not known exactly but 307 newtons at 5.00 ms^{-1} is not likely to be more than 10% from the truth.

If the crew of the trieres could develop 192 watts each, the total mean effective power would be 32.6 kw. According to the tank tests this would be sufficient, if the rudders were partly raised, to increase the top speed to 9.9 knots: see Figure 38.

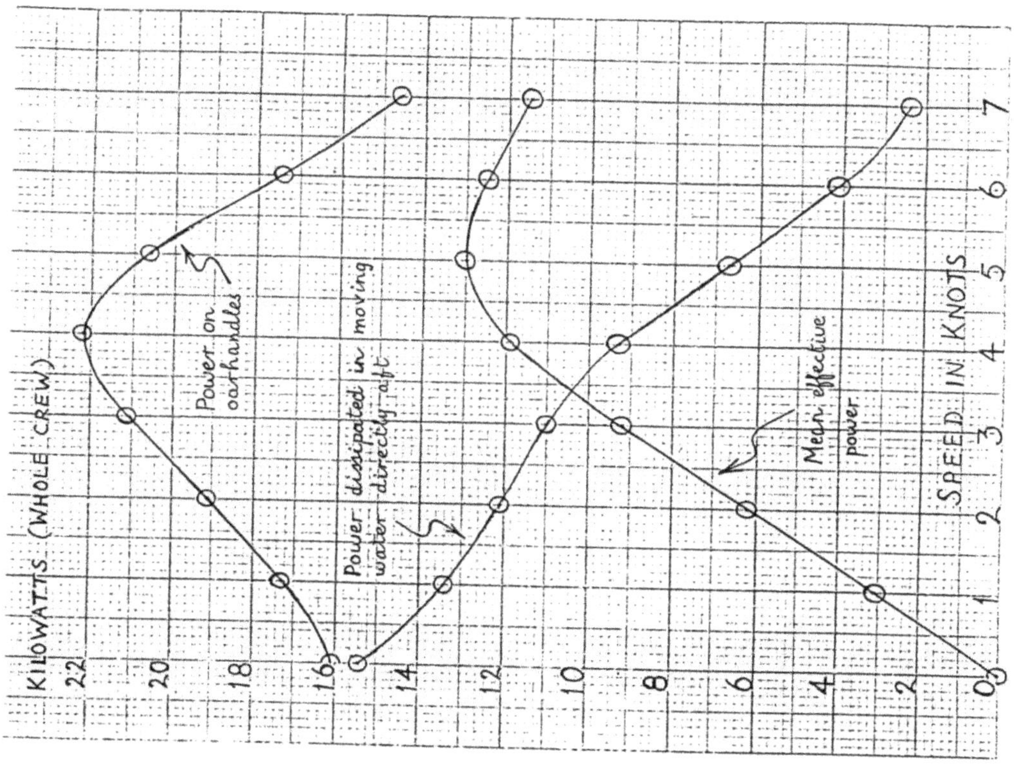

Figure 41 Power - Speed Relationship during Acceleration Trial

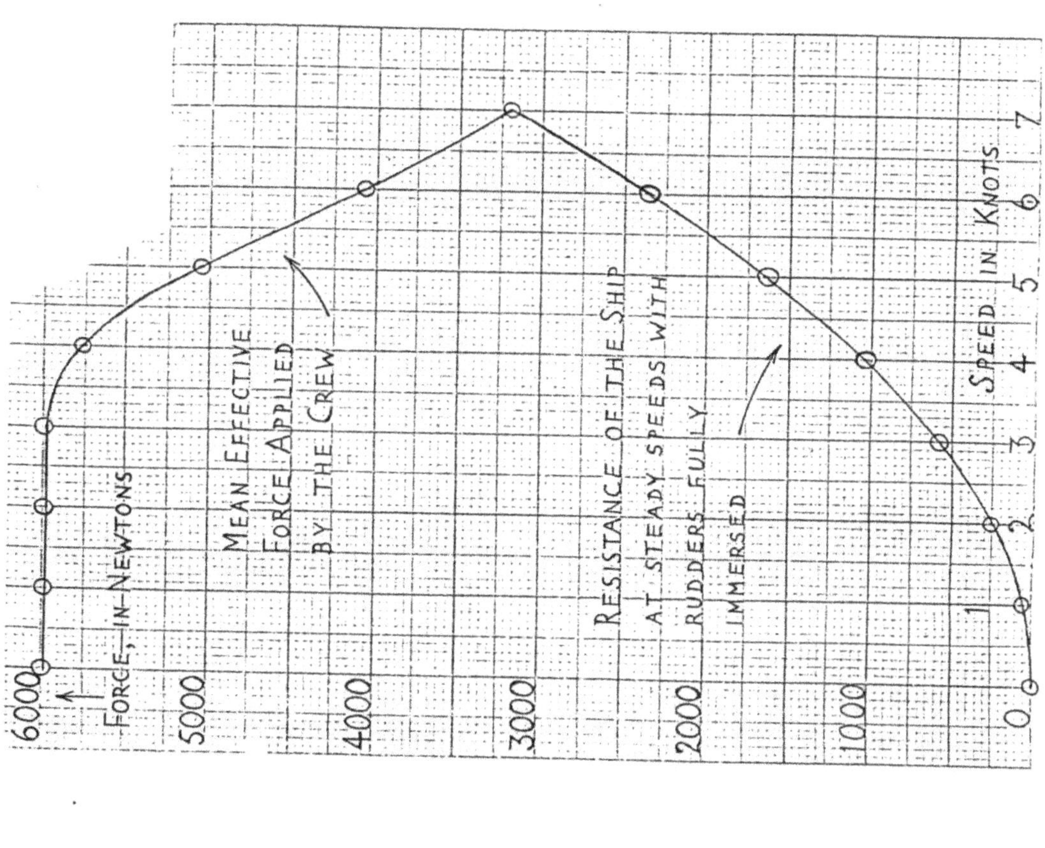

Figure 40 Mean Effective Force - Speed

ANNEXES B ROWING THE SHIP

There are probably several reasons why the power output of the crew of the trieres was comparatively low in 1987. They include:-

1. Unfamiliarity with fixed-seat rowing;
2. Cramped 'room' (i.e.rowers too tall);
3. Some of the crew not strong enough.
4. High weight and moment of inertia of the oars
5. Inequality of the oars

The first four of these would contribute to shortness of stroke, and all would engender a weakness of pull. It is the opinion of J.F.Coates that the average length of stroke by the crew in 1987 - i.e. the horizontal distance through which the ends of the oarhandles were moved relative to the structure of the ship whilst the blades were working in the water - was only 0.65m, whereas 0.80m or even 0.85m should have been possible. (Even 0.85m is much less than the corresponding distance in a fixed-seat eight which was about 1.22m at best; however the oars of the eight would have had a less severe gearing than those of the trieres in 1987.)

7.6. BLADE REACTION, PROPULSIVE FORCE, RHYTHM FACTOR AND RESISTANCE OF THE SHIP AT CONSTANT SPEED.

The blade reaction of a single blade is the force the blade exerts through the centre of pressure on the water. Necessarily, it acts at a right angle to the blade but it can be resolved into two horizontal components, one of which, at a right angle to the line of advance of the ship represents wasted effort, while the other, whose average value through the stroke is here called B, is parallel to the line , and gives rise to a reaction at the tholepin that is numerically equal to B. It is this reaction, multiplied by n, the number of oars, that propels the ship. Now when the speed of the ship, averaged over a few complete strokes, is constant, the propulsive effect of the intermittent force nB must be equal to the retarding effect of the ship's resistance, which is known: call it b. If now a rhythm factor f be defined, such that:-

$$f = \frac{\text{Duration of stroke plus run}}{\text{Duration of stroke}}$$

it becomes clear that :-

$$nB = fB$$

That is, the intermittent propulsive force exerted on the ship by the crew equals the product of the rhythm factor and the resistance of the ship, as long as the speed averaged over a few complete strokes is constant.

The rhythm factor is seldom measured. However, during the trials at Poros in 1987 a videotape was made that shows it. Its value during an acceleration trial is calculated theoretically in section 7 below, and the result is compared with the measurements from the videotape.

The blade reaction can be obtained from B if the angle θ between the oar and the line of advance is specified. Of course θ increases throughout the stroke, but it is possible to obtain a mean value for sin θ, knowing the length of the stroke, and so to calculate the mean blade reaction as B/sin θ, i.e. fb/n sin θ.

From the blade reaction, the dimensions of the oar, and the angle $\theta+\gamma$, where γ is the angle between the midline of the rower's arms and the line of advance, the magnitude of the pull on the oarhandle can be calculated. If the component of the blade reaction parallel to the line of advance is again called B and the distances from the tholepin to the centre of pressure of the blade and to the midpoint between the rower's hands are c and d respectively, then the pull on the handle is given by

$$\text{Pull} = Bc/[d \sin\theta \sin(\theta+\psi)]$$

$$\text{i.e. Pull} = fbc/[nd \sin\theta \sin(\theta+\psi)]$$

With the oars used in the trieres in 1987, if the stroke of the end of the oarhandle was 0.65 m, the mean value of $\sin\theta \sin(\theta+\psi)$ was about 0.956. The values of c and d were 2.90 m and 0.82 m..

7.7. THE RHYTHM FACTOR, THE RATE OF STRIKING AND THE MEAN SLIP OF THE BLADES DURING ACCELERATION TRIAL

This investigation is offered as a first approximation to the inter-relationship between the quantities listed below:-

f the rhythm factor

r the rate of striking

b the resistance of the ship, which is equal to the component of all the blade reactions measured parallel to the line of advance and averaged over stroke plus run i.e., as already shown, nB = fb

(= f x mean effective pull by the crew)

s the mean slip of the centre of pressure of all the blades measured parallel to the line of advance

M the mass of water accelerated by an average blade

a the mean sternwards acceleration of mass M

v the velocity of the ship

Regarding f:

It is probable that during the acceleration trial the crew were rowing as hard as they could. In that case their mean pull on their oarhandles while the blades were in the water would have been kept as high as possible for as many strokes as possible, i.e., it would have been constant at first. Since the mean effective pull averaged over stroke plus run has been shown to have been constant at first (up to 3 knots), it follows that the rhythm factor f was also constant within that speed range. Here, it is suggested that it would have remained constant throughout an acceleration up to 7 knots.

Regarding r:

Ian Whitehead's report (see above Chapter 2.3.4) states that during the acceleration trial the rate of striking began at 26 and rose to 42. These rates are averages over several strokes. The true rate during the first complete stroke is assumed to have been 25 s.p.m. and the rate at 6 knots 40 s.p.m.

The duration of a stroke in the water is 60/fr seconds.

Regarding s:

It can be shown from the dimensions of the oars that if a stroke is rowed with the ship held stationary, the slip of the centre of pressure of the blade (0.30m from the tip) is 1.848m if the end of the handle is moved through 0.65m.

Regarding M:

It is assumed here that the average working blade grasps, as it were, a certain mass of water M irrespective of the rate of striking and ship velocity and that it is the acceleration of this mass that generates blade reaction.

Resistances caused by shear are regarded as negligible in comparison.

ANNEXES B ROWING THE SHIP

Regarding a:

As a simplifying assumption the sternwards acceleration, a, of the mass M of the water grasped by the blade is regarded as uniform. Also, to help in keeping the argument simple it is assumed that the acceleration of the ship is uniform throughout any given stroke. This is not strictly true, but the errors are small. The effects of the longitudinal accelerations of the crew and oars relative to the ship are neglected because they very nearly cancel each other. The inboard stroke length is taken to be 0.65 m throughout the acceleration trial, and the rhythm factor is assumed to be constant also.

Consider the start from rest, i.e. the first stroke. The effective pull is 5957f newtons, the mass is 45 tonnes and accordingly the acceleration is 0.133f ms^{-1}. If the rate of striking is 25 s.p.m., the duration of the stroke, t, is 60/25f seconds. The distance moved by the ship in that time is :-

$0.5 \times 0.133f \times (60/25f)^2$ i.e. 0.383/f metres. The mean slip s_1 of the centres of pressure of the blades is that given by

$$s_1 = 1.848 - 0.383/f$$

During any stroke the acceleration a along the line of advance of the mass of the water M grasped by each blade is given by

$$s = ½\ a\ t^2 \text{ where } t = 60/rf$$

Therefore in general

$$s = 10.59\ b/fMr^2$$

The values of b and r during the first stroke being 5987 and 25 respectively it follows that

$$s_1 = 101.44/fM$$

That is

$$101.44/fM = 1{,}848 - 0.383/f$$

This result is used farther on.

Now consider the situation when the speed is rising through 6 knots. For convenience, assume there is a stroke for which the mean speed is exactly 6 knots. At 40 s.p.m. its duration is 1.5/f seconds during which the ship moves a distance of

$6 \times 0.5148 \times 1.5/f$ i.e. 4.666/f metres.

But this is equal to 1.848 metres minus the slip s_6 in metres.

At six knots the value of b is 4057 newtons. Little error is involved in assuming that this value holds throughout the stroke.

Putting b = 4057 and r = 40 into the expression

$$s = 10.59\ b\ /\ fMr^2$$

yields that

$$s_6 == 26.85\ /\ fM$$

Therefore $1.848 - 26.85/fM = 4.633/f$

Combining this with the equation found earlier, namely

$$1.848 - 0.383/f = 101.44/fM$$

yields that

$74.59/fM = 4.250/f$, i.e. M = 17.55 kg.

This is the mass of seawater accelerated by each blade and it amounts to 17 litres if the specific gravity of the water is 1.03. In relation to the dimensions of the blades the figure seems reasonable.

It is now possible to obtain values for f,s_6 and s_6 and s_1.

Since $s_6 = 26.85/fM$, $s_6 = 1.530/f$

But $4.633/f = 1.848 - s_6$ i.e. $1.848 - 1.530/f$.

Therefore f = 3.335 (say 3.3) and s_6 = 0.459 m.

Also s_1 = 1.733 metres. Say s_6 = 0.46m, s_1 = 1.73m.

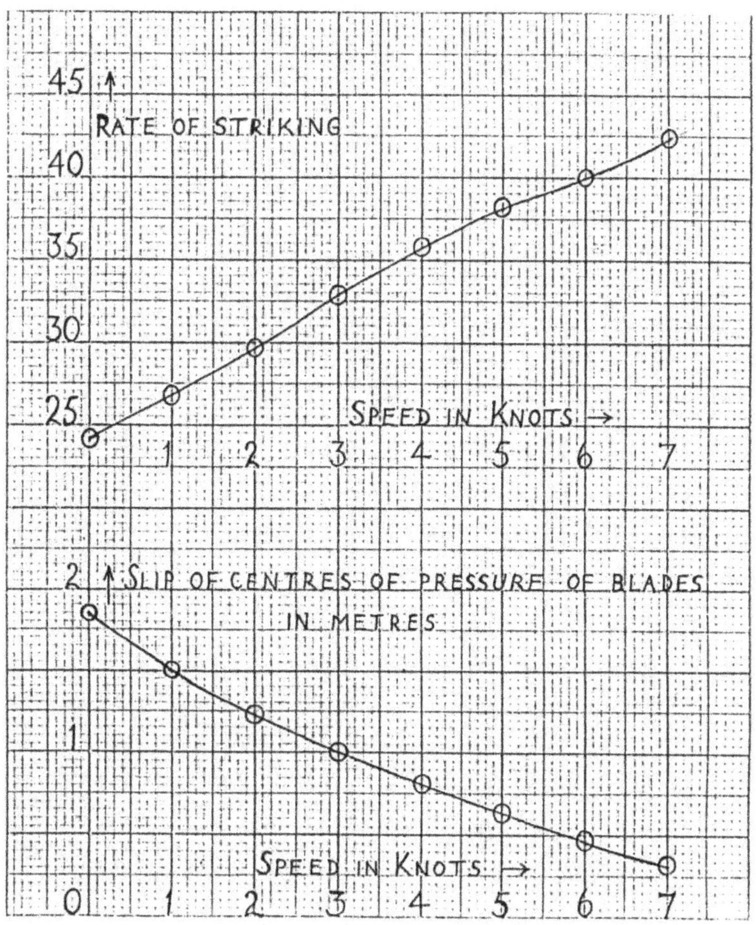

Figure 42 Slip and Rate of Striking against Speed

As a check on the value for f, a videotape showing the crew rowing hard at 44 s.p.m. was studied. The analysis was difficult because the bladework was not uniform. The best estimate of f was that it lay between about 3.4 and 3.6. This estimate seems to vindicate the foregoing method of calculation of f.

It is now possible to evaluate the rate of striking and the slip at all the speeds during the acceleration, including the speed of 7 knots that would eventually have been reached (it was reached in a sprint trial). The equations used are : $sr^2 = 0.1809b$

and $17.99v/r = 1.848 - s$

which yield the quadratic in r

$1.848r^2 - 17.99vr - 0.1809b = 0$

This can be solved after substituting for v in ms^{-1} and b. The results are tabulated below and are shown also in Figure 42.

Speed (v) knots	0	1	2	3	4	5	6	7
Rate of striking(r)	24.2	26.8	29.7	32.9	35.8	38.1	40.0	42.4
Slip(s)metres	1.85	1.50	1.23	1.00	0.81	0.63	0.46	0.32

In future tests an attempt should be made to measure these quantities as a further check on the method of calculation.

Similar details for the Oxford Etonians' eight of 1870 are unknown,'but not beyond all conjecture'. It can be said that a good fixed-seat eight of their period would have expected to average about 40 strokes per minute over the Henley course, and their rhythm factor would not be far from 4. From these figures and from the dimensions of the oars and the possible reach given in section 3 above a slip of 0.845m per stroke can be deduced. This is 2.77 ft and is possibly a little on the high side, to go by observation of modern crews (making allowance for the use of slides). If the rhythm factor had been 3.7 or the rate of striking 37, the slip would have come to 0.693m., 2.27 ft., which is perhaps about right. If this latter figure is accepted, then the mass of water 'grasped' by the blade is given by:

$0.693 = 0.5 \ (307 \times 3.7/8)(1/M)(0.405)^2$

i.e. M = 16.8kg (rhythm factor 3.7, rate of striking 40).

The quantity found for the trieres was 17.6kg. These figures are gratifyingly similar.

7.8. POWER EXERTED ON THE OARHANDLES AND POWER EXERTED WHEN MOVING WATER AFT These are not the same as each other or as the mean effective power that was dealt with in section 5. The latter is the power that goes to propelling and accelerating the ship. In order to exert it however the crew are obliged to shove water aft and also sideways, and these actions also absorb power, which has to be provided by the crew.

The power involved in moving water sideways is not calculated separately here but is included in the power exerted on the oarhandles. The power involved in moving water aft is also included here but it is of interest to calculate it separately. It is given by:

slip x mean blade reaction parallel to the line of advance x rate of striking/60 = watts.

ANNEXES B ROWING THE SHIP

The power exerted on the oarhandles is given by:
Distance moved by midpoint between hands x mean pull on handle x rate of striking/60 (=watts).

These quantities are tabulated below. The mean effective power is also shown (See also Figure 40).The powers are the averages over the stroke plus run. One sink of wasted power has been neglected as there is no simple way to calculate it: this is the power involved in creating turbulence in the water round the blades. It is small, but would have to be provided by the crew. Leaving it out of account, the percentage rowing efficiency, defined as:-

$$\frac{100 \times \text{mean effective power averaged over stroke plus run}}{\text{Power on oarhandles averaged over stroke plus run}},$$

can now be stated. It too is tabulated below:-

TABLE OF POWER (AVERAGED OVER STROKE PLUS RUN) AND EFFICIENCY DURING ACCELERATIOIN

Speed in knots.	0	1	2	3	4	5	6	7
Power on oarhandles, whole crew, kw	16.1	17.3	19.1	21.1	22.2	20.6	17.4	14.6
Power on oarhandles per oar, watts	95	102	113	124	131	121	103	86
Power absorbed in moving water aft,whole crew, kw	15.4	13.4	12.1	11.0	9.3	6.7	4.11	2.4
Mean effective power, whole crew, kw	0	3.1	6.2	9.2	11.9	13.0	12.5	11.5
Mean effective power, per oar, watts	0	18	36	54	70	76	74	68
Rowing efficiency, %	0	17.8	32.2	43.7	53.5	63,0	71.9	79.0

The power wasted in pushing water sideways is not tabulated but can be calculated as either the difference between the power on the oarhandles and the sum of the mean effective power and the power absorbed in moving water directly aft, or (which yields the same answer) as (1-0.956) times the power on the oarhandles. The figure 0.956 applies when the stroke of the end of the oarhandles is 0.65m. The power employed by the Oxford Etonians' 1870 crew in shoving water directly aft when travelling at 9.72 knots would have been about:

$0.693 \times 307 \times 3.7 \times 40/60000 = 0.52$ kw (averaged over stroke plus run)

on the bases that the slip was 0.693m, the mean effective pull 307 newtons, the rhythm factor 3.7 and the rate of striking 40.

The power they exerted on the oarhandles would have been about:

$$0.99 \times 307 \times 3.7 \times (2.32/0.84)(1/0.882)(40/60000) = 2.35 \text{ kw}$$

i.e. 293 watts per oarsman, 0.393 hp averaged over stroke plus run; on the bases that the midpoint between the hands was pulled through 0.99m, the oar dimensions were as given in section 3, and that $\sin\theta \sin(\theta+\psi)$ averaged 0.882 for the given length of stroke with the givenoars. The effective mean power (i.e. effective in propelling the ship), as already pointed out, was about 192 watts per oarsman, 0.257 hp; or 1.54 kw for the whole crew.

7.9. RATE OF STRIKING AND SLIP WHEN PADDLING AT STEADY SPEED

The Newsletter report on the 1987 trials states that 5 knots were attained at 30 spm and at light pressure. From the tank tests the mean effective pull was then $65.2v^2$ (v in knots) = 1630 newtons.

Equations for s and f are then:

$$s = 1.093/f$$

$$5.148/f = 1.848-s$$

from which $f = 3.377$ say 3.4

and $s = 0.324$ m. i.e.,1.06 ft.

Analysis of the videotape suggested that f was about 3.7 when the rate of striking was 30, but it was not certain that the crew were paddling light when they were videoed at 30 s.p.m.. They may have been going harder , which could account for a higher value of f.

7.10. WAYS OF IMPROVING THE SPEED AND ACCELERATION OF THE TRIERES

These are mostly rather obvious, and some can merely be listed without discussion. They are:-

1. Obtain more practice in fixed-seat rowing, preferably in the trieres.
2. Select smaller but stronger people than in 1987.
3. Lighten the oars and improve them in various details (e.g.,the length of the hand-grips).
4. Improve the oarloops so as to minimise the loss of rowing length caused by their slackness.

By these means it should be possible to improve the performance of the present trieres considerably. There are however three further points:

5. Because the trieres floated too high, making it hard to cover the blades, about 5 tons of ballast were put aboard. The ability of the crew to exert power may have improved as a result, but the mass to be accelerated and the ship resistance both increased. This is not the place to go into the consequences, but it is suggested that consideration might be given to taking the ballast out again, and even lightening the ship yet further by removing the masts and spars. Adjustments to the heights of the seats and tholes would be needed. (Admittedly this would be difficult in the present ship).

6. The rudders and their stocks cause a high resistance that can be much reduced by lifting the rudders partly out of the water. Arrangements to facilitate this operation should be made.

7. When the design of the trieres is reconsidered, every effort should be made to find a way of allowing the thalamians to row a longer stroke in the confidence that they will not be hurt by the transverse beams at present in their way.

7.11.ATTAINMENT OF 9.5 KNOTS IN THE PRESENT TRIERES

For this calculation it is assumed that the oars will not be modified in time for the 1988 season. The required mean effective power (from tank tests) with rudders partly raised will be 26.8 kw and the mean effective pull 5490 newtons. Then, if the crew has been persuaded to row a full stroke, 0.80m at the handle-ends, the slip s will be given by:

$4.89 = (2.275-s)fr/60$. But s is also given by:

$$s = 0.5 \ (5490f/170)(1/17.55)3600/f^2r^2$$

If f is assumed equal to the value 3.335 found previously, these equations yield a quadratic in r the solution of which is r (i.e. the needed rate of striking) = 47.8,and therefore s the slip = 0.43m.. Although a mean effective power of 26.8 kw, i.e., about 158 watts (0.211 hp) per oar, should be attainable by a fit crew for long enough to demonstrate a speed of 9.5

knots, great skill will be needed to row at 47.8 s.p.m. and at the required length.

For a stroke of 0.8m at the end of the oarhandle, sin θ sin $(\theta+\psi)$ averages 0.925. The pull on each oarhandle averaged over the stroke would have to be 412 newtons, 42.0 kgf or 92.5 lbf. The power exerted on each oarhandle would be 211 watts, i.e., 0.283 hp.

7.12.CONCLUSION

Data from the trials in 1987 and from tank tests of a model have been used to show what forces and powers were generated by the British crew of the trieres when they carried out the test of acceleration. Rowing in a fixed-seat racing eight has also been investigated in order to show what power might be expected from a suitably selected and trained crew of a trieres.

The mean effective power (m.e.p.) per oar of the 1987 trieres at her top speed of 7 knots was just under 68 watts whereas in the eight that was investigated the m.e.p. per oar was about 190 watts at the higher speed of 9.7 knots.

This annex contains an analysis of the rowing of the trieres, covering rates of striking and slip of blades as well as forces and powers exerted. This leads to the suggestion that to increase the top speed to 9.5 knots it will be necessary to increase the length of stroke inboard to 0.8m and to raise the rate of striking to over 47. Clearly this will require great skill and determination, qualities which were possessed in abundance by the Athenian oarsmen of 480 BC.

ANNEXES B ROWING THE SHIP

B.8 HOW TO ROW A TRIERES: S.F.Weiskittel

8.1. ORGANISATION OF THE OARSYSTEM The oars of a trieres were not ranged along the sides of the ship at equal intervals but grouped in threes, each triad consisting of a thalamian, a zygian and a thranite. This arrangement was necessary to ensure that the three-dimensional volume swept out by each shaft during the course of a stroke would remain clear of obstructions, whether of the other oars or of structural elements of the ship or of the bodies of the rowers themselves. Even so, this was barely accomplished in parts of the ship. For example, at the extreme bow, the shaft of the zygian oar during the drive phase of the stroke would ride along the underside of the horizontal beam supporting the seat of the thranite above, and sometimes even the calf of the thranite above, potentially limiting the depth of the blade in the water. Rowing in such a position an oarsman could not tell if his blade was fully immersed, especially on occasions when the ship heeled over to the other side.

This triadic arrangement seems plausible on historical grounds, assuming that the trieres evolved from a two-level pentecontor. This earlier ship, herself evolved from a single-level pentecontor, would have spaced her oars at equal intervals along the side, alternating thalamian and zygian blades. The introduction of a third level above was accommodated by adding a third oar between each thalamian/zygian pair. Since each triad of blades shared in effect the same small stretch of water, however, perfect timing was necessary to avoid clashing, and there was considerable turbulence of the water, lessening the efficiency of each oar.

There was much discussion among the crew regarding the effectiveness of this arrangement, especially with regard to the churning of the water. At least one experiment was undertaken to see if it would be possible to decrease this turbulence. It was noted that the Lenormant relief showed the thranite tholepin situated nearly vertically above the thalamian. For the reconstructed ship, this was rejected as impossible, since a design based on the Lenormant relief would find it difficult to support the thranite seat in this location without simultaneously interfering with the zygian oar volume. Thus, the Lenormant relief was interpreted as showing not a broadside view, as commonly understood, but a view from 70-75 degrees off the bow, with a consequent shifting of the thranite pin towards the bow. For a discussion and illustrations the reader is referred to J.F.Coates in *AT* Appendix II, pp.234-237.

If the thranite tholepin were situated above the thalamian, however, and if it were possible to set the thranite slightly further out from the centre line of the ship, a radically different oarsystem would result, with the thranite and thalamian blades forming a single entity, one blade reaching out slightly further than the other, sweeping together in the same vertical plane, followed by the zygian blade, with a considerable reduction in the immediate water turbulence. To test this possibility in the reconstructed ship, a thalamian blade was pulled in slightly (since it was impossible to set the thranite blade out further), and thranite and thalamian blades were rowed together in front of the zygian. The thalamian blades generally, in any case, were pulled in slightly during rowing (see below section 7). It was certainly possible to row this way, though it was not clear whether the results justified the effort. The heavy oars meant that there was very little feel for the water, and thus it was not possible to decide whether the slight reduction in turbulence permitted a more effective stroke.

8.2. EFFECTIVENESS OF THE OARSYSTEM The ship's speed was dependent on two variables, stroke rate and length of stroke. The crew seemed most comfortable rowing at a rate of thirty-two or thirty-four strokes per minute, though the rate rose above forty on occasion. It was clear however, that there was considerable variation of strength and fitness in the crew: outings that brought some rowers to the brink of collapse were considered reasonably easy by others. It is likely that a crew chosen more carefully for strength and fitness could row at higher rates comfortably, and this ought to be a matter for further experimentation. In 1987, high rates threatened the safety of the thalamians because of the danger of catching a crab (see below).

Maximum stroke length was rarely achieved in 1987. Most of the rowers on all levels rowed relatively upright, taking the stroke with arms extended, touching or nearly touching the back of the rower ahead. The taller members of the crew with long arms could swing very little from the hips; shorter members with shorter arms could swing rather more: and in general the rowers felt most comfortable rowing in a style roughly similar to what they were accustomed to use, i.e., sliding-seat rowing, with arms away at the finish followed by as much body swing as possible.

It was apparent that this style needed adaptation for the fixed-seat rowing of the trieres. Because of the cramped conditions of the ship, maximum length of stroke depended on maximum reach at the catch. The catch, however, was limited by the back of the rower in front (i.e., towards the stern which the rowers faced), and a longer stroke could only be achieved by having that back move further away so that the handle of the oar behind could be pushed further forward. Accordingly, maximising stroke length required maximising body swing. Some of the rowers increased their body swing by rowing with bent arms. Although this technique in itself did not lengthen their own stroke, since their length was still limited by the degree to which the rower in front swung forward, their greater lean forward did allow the rower behind to increase his stroke length. Unfortunately, the swing of backs of a group of rowers rowing like this was often interrupted by a single upright rower behind the group. Given the general size of the crew, and their long arms, some bending of arms at the catch was inevitable, even with maximum body swing of the rower in front, but a smaller crew ought to be able to row both with arms extended and maximum body swing.

The maximum potential body swing of the thranites and zygians was considerably greater than that of the thalamians. The thranites and zygians had no structural limitations on their movements and could bend forward considerably more than the thalamians, whose forward and backward movement was limited by the thwarts or main transverse beams which passed between each thalamian and the next at head level. Full body swing for a thalamian accordingly required great self-confidence and control, since at the catch the thalamian's nose would touch the beam in front, while at the finish the back of the neck would graze the beam behind. Therefore,since the movement of the thalamians was more constrained than either the thranites or zygians, the length of stroke for the whole ship had to be set by the length of the thalamians' stroke. Unless they were coached not to do so, the zygians in particular tended to row a longer stroke than the thalamians, thus interfering with the rhythm of the triad.

Coupled with the general upright style was a tendency to use the arms to pull the oar through the water, especially with the stronger rowers. This was much less effective than using the back and body weight to move the oar. The cramped position of rowers rowing with bent arms at least had the great benefit of requiring the force of the stroke to be taken by the back rather than the arms, giving a sharper and quicker catch. Even so, there was a general lack of sharpness at the catch, and not much feeling of moving the ship during the drive. Part of this was due to the considerable turbulence of the water, churned up by the oars, part to the general heaviness and stiffness of the oars. It is likely that lessening the weight of the oars and increasing their flexibility will provide more feeling of the weight of the water at the catch and during the drive.

The oars were tied to the tholepins by leather loops so that one rowed against the loop rather than against the pin. There was considerable disagreement among the crew as to whether this was better than rowing against the pin. During each stroke, the loop stretched, so that at the conclusion of the stroke, the horizontal vector of the force required to lift the oar from the water caused the oar to snap back against the pin. This constant hammering caused nearly all the tholepins to wear in their sockets, becoming loose and frequently popping out. Since different loops stretched to different lengths, it also meant that rowers finished the stroke at slightly different times, and the constant vibration transmitted to the hands at each stroke caused considerable numbness after ten minutes or so of continuous rowing. This latter problem was most severe for the thranites, less so for the zygians, and barely noticeable for the thalamians: the higher the level, the more force was needed to lift the oar from the water, and accordingly the greater the horizontal vector causing the oar to snap back against the pin. The constant stretching and hammering caused oarloops to wear excessively and break frequently. As the supply of loops was limited, a number of rope grommets was made to replace them. These stretched less and were generally preferred to the leather loops.

One potential advantage of rowing against a flexible loop rather than against a pin was to avoid variations in the pitch of the oar. Although in general there was no difficulty in keeping the oar square to the water, many of the thalamians had problems in doing so. The difficulty seemed to be with the thalamian tholepin block, which interfered with keeping the blade square if it was pressed right out against the block. Many of the thalamians rowed with their oars pulled into the ship slightly, avoiding the problem and at the same time allowing a more comfortable position during the drive, allowing the hands to finish more towards the inboard side of the body than towards the centre.

Occasionally, the thole loop itself caused difficulties. Most of the crew did not appreciate that the starboard loops were constructed differently from the port loops, the one with a clockwise twist and the other with a counter-clockwise twist. If a port loop was fitted to a starboard oar or vice-versa, the loop itself would tighten on the oar as the handle was raised to put the blade in the water, thus binding the oar and holding it on an incorrect pitch.

8.3. ORGANISATION OF THE TRIAD The numbering of the seats in the trieres did not correspond correctly with the organisation of each triad. For example, the two positions coupled with thranite no.10 were zygian no.10 but thalamian no.11, that is, each thalamian seat was numbered according to the row with which it was most closely connected spatially in the ship, but not with that with which it was most closely connected organically, so to speak, so that each thalamian seat number was one more than its correct triad number. This discrepancy in the numbering caused considerable confusion during one outing when the bow sections were switched with the stern sections. The switch was done to compare the relative performance of the sections, since the bow section invariably generated faster ship speeds than the stern section. It was unclear whether this superiority was due to the individual rowers or to their position in the ship. A fair test, however, required keeping intact the triads, since the key to successful rowing was the integrity of each triad and the constant practice within a triad to achieve synchronisation. Since the bow and stern sections divided at positions 10 and 20 from the middle sections, which split up two triads, and since the first two triads were incomplete in the bow and the last three triads in the stern, it was impossible to simply switch the rowers between sections. This was not correctly appreciated, however, and a straight swap was attempted, with inconclusive results.

Instead of beginning at no.2 and ending at no.28 (Fig,33), the thalamians should have been numbered starting at no.1 and ending at no.27. This would clarify the proper position of the thalamians, especially when substitutions were necessary, when it was important to know to which triad a particular rower should belong.

There was considerable variation of technique among the rowers, some of which could be corrected, but some of which seems unavoidable without the constant months of practice recorded of ancient Athenian crews and which are likely to be impracticable for trieres crews. Even given such practice, minor differences are likely still to remain. These differences can be and were counteracted by effective rowing within each triad since the idiosyncrasies of each rower were accommodated by the other two rowers of the triad. Accordingly, it was essential that each triad was composed of the same rowers on each outing. Towards the end of the trials, when a number of the rowers had left Poros early and there were general substitutions all over the ship, there was deterioration of the rowing. It took more than one outing to become familiar with the other members of the triad.

8.4. TIMING AND COMMUNICATION Only the thranites could keep effective time along the length of the ship, because only they could see down the length of the ship to the stern. Even so, there were numerous occasions when the timing went off even with only the thranites rowing; and when this happened, it took several strokes for the correct timing to be reasserted. This was because it was necessary for each thranite to keep in time with the other thranites immediately around as well as be concerned with the overall timing. A timing correction would accordingly ripple down the ship as the rowers out of time and closest to the stern would correct their stroke, followed by those immediately behind,and so on.

Since only the thranite could see the oars, he needed to communicate information about timing and other difficulties to his zygian and thalamian. Often the thalamian or zygian knew something was wrong without being certain what it was, and asked for information. Communication between thranite anbd zygian was easy, since they were sitting so close. The contrary was the case for thranite and and thalamian. Usually neither could clearly hear the other, and information had to be relayed up or down by the zygian.

There was considerable criticism from the Rowing Master and the section leaders about the constant chatter in the ship which made hearing of commands difficult, but much of this chatter was the important relaying of information about the triad from the thranite or to the thranite. Triads with constant membership needed considerably less information than newly formed ones, and no doubt more practice and greater consistency in triads would serve to reduce the need for talking in the ship.

The thalamians and zygians were entirely dependent on the amplified commands "in...out" or "there...there" for picking up the correct timing. It was necessary for the entire triad to alter its timing simultaneously, and this was difficult to do without some external audible mark. It was exceedingly difficult from the mainmast forward to hear commands from the stern. Even when a relay was stationed at the mainmast, it was difficult to hear in the extreme bow. Given literary references to timing chants and the presence of a piper in the ancient ship, it seems likely that the timing was taken from audible signals in antiquity.

8.5. THE STROKE : GENERAL CONSIDERATIONS Given the necessity of maximising length, one effective method was to touch or graze the back of the rower ahead when reaching forward for the catch. Many of the rowers were reluctant to do this, reflecting no doubt their experience rowing in sliding-seat boats, where the rapid leg-drive at the catch meant that contact with the oarhandle behind was quite painful, and they were afraid to chance inflicting this on the rower ahead. The motion backwards in the trieres was not so powerful, however, and any contact was with the hands on the handle rather than the handle itself, given the angle of the oar at the catch, so that there was no danger of injury, though poor timing could produce a minor bruise. This technique permitted the greatest reach forward.

The great majority of rowers sat on pieces of sheepskin. These were unnecessary for comfort, as the few rowers who did not use them will testify, but they did facilitate a slight sliding or rocking backwards on the buttocks at the catch for the shortest rowers. All of the rowers tensed their legs at the catch, transmitting the force of the handle to the footstretchers. Most of the rowers, because of their height, had to keep their knees bent to avoid pushing themselves backwards off their seats. The shortest rowers, however, who could row with legs fully extended, bent their knees slightly at the catch, then straightened them during the drive, rocking backwards and lengthening their stroke slightly. From time to time, however, they had to readjust themselves slightly forward on to the centre of the seat off which the cumulative series of strokes had moved them.

When a section of the crew rested while others pulled, it was useful to have some way to keep the blades clear of the water without having to pull in the oars. Because of the weight of the oars, it was awkward and uncomfortable simply to hold the handles down. The thalamians and zygians tied short pieces of rope to the supporting beams running between their legs, with loops at the ends, hooking these over the handles when they were resting. Provided that the length was adjusted correctly, this technique kept the oars clear of the water without the necessity of holding the handle down, which was tiring to do for very long. It also allowed the handle to be freed quickly when the order came for the thalamians and zygians to start rowing again. Some of the thranites tried the same device, but the correct length of rope to hold the blade out of the water was just slightly too short to allow the loop to be quickly and easily slid off the handle again. Thus, most of the thranites simply hooked the oar under their arm and used the weight of their upper bodies to keep the oar clear. This was possible for them because the greater shaft angle meant that the handle was still quite high in the ship when the oar was clear of the water. The thalamians and zygians could not comfortably do the same thing since their shaft angles were so much less.

When the order was given for a section of the crew to pull in their oars completely while on the water, the zygians were directly in the way of the thranite oars. The thranites accordingly lifted their shafts and rested the looms on stanchions supporting the upper canopy. The zygians pulled in their oars so that the looms rested on the gangway, but this was not entirely comfortable because the shafts would then rest on their thighs and trap their legs underneath. Some of the zygians would extricate their legs so that the shafts passed underneath. The thalamians simply pulled in their oars as if for disembarkation and could then stand up on their seats to stretch their legs.

8.6.THE THRANITE STROKE The important role of the thranite in the organisation of the rowing has already been mentioned. Within the triad the thranite oar occupied the space and the water between the zygian and the thalamian oars. This space could increase or decrease the differences in timing between the zygian and thalamian. Such differences could only be corrected with observation and comment by the thranite, who could see where the problems lay. Often the gap between the two other oars would completely disappear at the catch if the thalamian oar moved further back towards the bow than the zygian.(The same of course would be true at the finish.) It was especially important in this case for the thranite to keep his blade as close to the water as possible during the recovery, so that the thalamian oar could not pass beneath and be caught by the thranite shaft. Since the thranite oar was especially heavy to manipulate, the general tendency was to heave the handle down at the finish to get it out of the water, raising the blade too far from the water and allowing the thalamian blade to pass underneath. It took considerable practice and arm strength to minimise this motion.

Sometimes this skying of the thranite blade was caused by the zygian lifting his blade too far from the water during the recovery. Because the two oar shafts were very close during the stroke, if the zygian blade was lifted too far from the water, the zygian shaft would catch the thranite shaft from beneath, and the thranite would have to lower his hands to clear hgis shaft. At this point the thalamian blade would sweep underneath.

At the finish the thranite had to clear his oar from the water in unison with the thalamian. If he did not, the thalamian oar would immediately swing over the thranite oar, so that the two shafts crossed and the blades became entangled. If the thranite and thalamian blades moved together but the zygian was late at the finish, the thranite oar would swing over the zygian oar, so that these two other shafts crossed and the blades became entangled.

Most thranites rowed with the palms outward and fingers resting on top of the handle as with a normal sweep oar, but the angle of the shaft was such that it was also possible to reverse the inboard hand with palm facing towards the rower, as practised by the thranite oarsmen in the trieres on Trajan's column (*AT* fig.4). This practice made it somewhat easier to lever the oar down out of the water, and helped to control the blade height during the recovery phase of the stroke. With both hands above there was a tendency to pull down too far as the rower tired, skying the blade, potentially causing the difficulties with crossing shafts already mentioned. It was often comfortable to alternate styles, rowing a few minutes one way, then changing to the other style for a few minutes.

The thalamian and zygian rowers were clearly visible from the thranite position, but it was much more useful to keep watch closely on the movement of the blades in the water, to anticipate developing problems. Many thranites would also alert the triad to approaching swells caused by passing cruise ships or ferries, but it was not clear that this was necessary. Usually advance warning did not prevent the synchronisation of the triad from deteriorating, and usually an experienced triad would correct the timing within one or two strokes without difficulty once the swell had passed.

8.7. THE THALAMIAN STROKE The thalamians had by far the easiest rowing, except for the dangers of catching a crab, though their location at the bottom of the ship had its drawbacks. For one, it was impossible to see out of the ship at all while rowing, though there were some very good views of the sky and the gangway and of the Captain and Helmsman at the stern. Secondly there was no breeze, and a constant and annoying drip of perspiration from the rowers above. On the other hand, when the thalamian oars were pulled in, the thalamians could get up and stand on their seats, and from there they had a very good view both out of the ship and up along the central gangway, and they also had an opportunity to stretch their legs, which the zygians and thranites could not do.

Thalamian rowing was very similar to normal sweep rowing, the perceived oar weight and the angle with the water was close as well. One difficulty was with the tholepin blocks, which prevented the oar from being fully pressed out against the tholepin. Instead of the shoulder of the oar (the point of transition between the round oar shaft and the square counterweight) properly resting against the tholepin, it was usually at a point about seven centimetres further along the shaft because the edge of the thole block interfered with the square counterweight. Furthermore, often the oar would fail to remain square to the water as it developed its own natural pivot point, and it was quite fatiguing to have to try to hold it square for each stroke.

Some of the thalamians tied a sheepskin around the thwart in front to protect their heads as they attempted to swing as far forward as possible for the catch. Others found it useful for gauging the distance to have the zygian in the next triad place his hand on the beam when he was not rowing, so that it was his hand rather than the wood of the beam which was touched in the swing forward.

The main, often harrowing, drawback in rowing thalamian was catching a crab. Because of the ease with which the blade could normally be extracted from the water, a thalamian caught a crab only when there was interference with his stroke from the other oars. The great danger was for a thalamian to come out at the finish marginally later than the zygian sitting above and in front, e.g., the zygian in the next triad towards the stern. This would mean that the zygian shaft would cross over the thalamian shaft before the thalamian blade was fully clear of the water. The zygian blade would prevent the thalamian blade from passing, and at the next catch would drive the thalamian blade down early and pull it though the water. Even though the zygian blade would then come out in a normal finish and release the trapped blade, by then it was too late; the motion of the ship and the pressure of the water would carry the oarhandle uncontrollably backwards, pinning the neck of the unfortunate thalamian against the back thwart. The Barrack Master, Peter MacLeod, was stationed in the bilges with a whistle, which he would blow when this happened, at which signal everyone would immediately stop rowing. This whistle was clearly audible and very effective at stopping the rowing. MacLeod also attempted to get to the distressed rower to help pull the shaft away, but the difficulty of clambering over the bilges meant that unless he was quite close and on the correct side of the mainmast support, it took too long to reach the thalamian. A more effective means of rescue was for the zygian above and in the same triad to stop rowing, place his foot on the thalamian shaft just below, and push it away. As rowing skill improved, whistle stops became fewer, though disaster lurked whenever the crew was rowing at full pressure.

To avoid such problems it was particularly essential for the thalamian to keep a close eye on, and stay in time with, the zygian above in the next triad sternward. One side effect was a reluctance on the part of some thalamians fully to pull in at the finish, the most dangerous point of the stroke, for fear of not getting the blades out of the water quickly enough to stay in time with the sternward zygians. Better to be slightly early, they reasoned, than to be slightly late. Since the thalamians must feel secure enough to pull in fully at the finish, some method needed to be found to avoid problems when thalamian and zygian blades crossed, as must sometimes have happened even in the best crews in antiquity. With coaching it should be possible for the zygians to identify the feeling of the oars when they are entangled with the thalamians, a feeling which the thalamians were quite well acquainted with, and to practise the correct response,which is to lighten up the stroke, not pulling though the water but just going though the motions, until the blades are clear again at the end of the stroke cycle.

8.8. THE ZYGIAN STROKE The perceived weight of the zygian oar was less than the thranite's because the angle with the water was less, but it was still considerable. The zygian had to stay completely in time with the thalamian in his triad to leave the correct interval between blades for the thranite. Since the thalamian sat below and forward of the zygian and so was completely visible, one might think that it would be easy to keep in time, but it was still difficult to follow his stroke. If the zygian tried to watch the thalamian's hands, he was invariably late in the finish. The thalamians all accustomed to sliding-seat rowing, executed a rapid hands down and away at the finish, just as if they were in a racing eight. Because the oars felt more or less the same as in an eight, this was a natural movement, and there was also the urgent need to avoid being later than the zygian in front, already mentioned, because of the dangers of catching a crab. The zygian in the same triad, however, was manipulating a heavier oar (not in fact, of course, but in perception) and because of the angle had a slightly greater arc to achieve to free the blade from the water. Thus, even if the hand movement downwards was commenced at exactly the same moment as the thalamian's, the thalamian would get the oar down and away faster towards the bow, and the gap between the thalamian and zygian oars would decrease, causing difficulties for the thranite, who could not follow the thalamian without hitting the thalamian blade, or worse, crossing the zygian shaft if the zygian was late out of the water.

The best solution for the zygian was to watch the loom of the thalamian's oar rather than his hands, and match the movement of the loom away towards the stern, keeping the shafts parallel, even if this meant commencing the downward motion of the hands a little early. A sharp pull in at the finish also helped, allowing the momentum in to be converted into momentum out, permitting the zygian oar to match the quick hands away of the thalamian. Even better was to encourage the thalamian to pause momentarily after the hands down; but success in this depended upon the next zygian towards the stern not getting ahead.

8.9. BACKING WATER It was very awkward to attempt to back water by pushing the oar away, and the ship did not back very quickly. Part of the problem stemmed from the excessively heavy oars, and with lighter oars this manoeuvre might well prove easier. Certainly an ancient trieres must have been able to back water quite easily and quickly, especially after ramming another ship, when it would have been essential to back away to free the ram and to avoid being an easy target for an approaching enemy ship. Because it was comparatively easy to row the ship forward, the Rowing Master tried having the thranites and zygians spin round in their seats, each rower taking the oar of the rower immediately behind, and actually rowing the ship backwards without the participation of the thalamians. Although there was some awkwardness in finding an appropriate place to brace the feet, the arrangement worked quite well with only two levels.
(The absence of any ancient evidence for this procedure for backing at any period, the non-participation of the thalamians, the difficulty experienced by the participating oarsmen in bracing their feet, as well as the confusion such movements would cause in the heat of battle, make its use very unlikely. Ed.)

8.10. TRAINING AND COACHING CREW Because of the general unfamiliarity with the ship and uncertainty about the best way to row it, there was little done in 1987 systematically to coach or train the crew. The most important effort was to inform the thranites to keep their blades berween the thalamians' and zygians' to prevent their shafts crossing. There was also some rowing by individual levels and rowing by dividing the ship in thirds (bow, midships and stern). Rowing a single level at a time was only minimally useful, since the key to good rowing was meshing in with the rest of the triad. Rowing a third of the ship at a time, on the other hand, was quite useful, as it meant that any timing difficulties could be fairly quickly corrected, since timimg changes would not have to go down the ship very far.

There are several discrete areas to be considered in training a trieres crew to row correctly. First and most important is the necessity of training the crew to row in the correct style with maximum stroke length. Second is training the crew to row appropriately to their position, whether thalamian, zygian or thranite. Third is training the rowers to row within their triads correctly, meshing their blades with those of the other members of the triad. Training and coaching methods ought to be aimed at teaching proper technique in the most efficient way possible. Given the time constraints on training any particular crew, one principal aim ought to be to keep as many people rowing as possible at any one time, so as to maximise the time spent actually rowing, which is what the crew, paying their own way, has come out to do.

ANNEXES B ROWING THE SHIP

The first thing is to coach the rowers in the proper style. Sliding-seat rowers in particular will need re-training not to row with the curved back and pronounced forward reach so prized in sweep rowing, where the stroke can be lengthened by training the crew to reach out further. In the trieres the maximum length is instead set by the cramped physical conditions of the ship, so the aim should be to utilise the given maximum given maximum length as efficiently as possible.

The length of a stroke is the sum of two factors: the distance the oar handle moves from the finish position away from the body by extending the arms forward, and the additional distance gained by bending forward at the waist. Even a smaller crew will find that the sum total of these distances is greater than that possible to achieve in the trieres: all three levels are constrained by the rower in front, the thalamians in addition by the front thwart. The range of motion is limited, therefore, by the degree to which the rower in front (i.e.towards the stern) gets his back out of the way. That is, as much as possible of the length of the stroke should be generated by swinging forward from the waist, with whatever additional space left being given to the arm extension, to allow the rower behind as great a range of motion as possible. In addition, the back should be kept as straight as possible to limit the amount of stroke length generated by rounding the back, which would allow the oar handle to be pushed out further without concurrently allowing the oar handle of the rower behind the same degree of travel. The most important thing is to have each rower generate the same stroke length in the same way.

Most rowers will find that they will not have the room to extend their arms fully at the catch. The ship was designed for a maximum handle stroke of 85 cm at the thalamian position which, as has been already mentioned, sets the stroke length for the whole ship. The transverse thwarts at the thalamian position permit a maximum body swing of 45 cm or so, leaving only 40 cm for the arm extension. Most rowers accordingly will row with some degree of bending at the elbows; the longer their arms the more bending will be necessary.

The necessity of coaching the thalamians for maximum stroke length deserves particular attention. One way of encouraging this would be to have the zygians place their inboard hands on the thwarts,gripping the beam with their fingers towards the bow, and then have the thalamians row together, attemmpting to touch the zygians' fingers with their noses at the catch, and grazing the thwart behind them with the backs of their necks at the finish. This would produce maximum body swing, the rest of the stroke length made up by extending the arms until they touched the back of the rower in front. The dangers of catching a crab also would be avoided by not having the zygians rowing.

The thalamians could row by themselves for this exercise, but a more effective use of crew time would be to have the thranites row together with the thalamians, the thranites attempting to keep a constant distance between their blades and the thalamian blades. Since they can see both the thalamian rowers and the thalamian blades, they can work at achieving the same body angle as the thalamians, neither more nor less, while matching their stroke length in the water. An audible beat would bve useful to mark the catch and finish, since the rowers need fixed points to which to match their stroke. The thranites need to be coached not to row a longer stroke than the thalamians, which is possible for them, but to keep in time, pulling through the water at the same speed, and using the same amount of time for the recovery. It is very easy for the thranites to develop bad habits, since they alone are used to take the ship away from the quay and bring it back in, without having to synchronise with the thalamians.

The next step would be to have only the thalamians and zygians row together, this time coached by the thranites to keep their oars at a constant distance apart. The pull through the water must be at the same speed, the time spent in the recovery must be the same, and the shafts must be parallel at both the catch and the finish. These crucial points must all be familiar to the thranites by this time, since they have had to try to match these elements themselves, and they should be effective in coaching the zygians. By isolating the top two levels of rowers in this way, each can learn how to maximise its own length of stroke and keep in time with the thalamians, while the thranites can concentrate on understanding the stroke and build confidence in their own coaching abilities.

The thranites and zygians should also practise keeping their blades close to the water on recovery. This will be crucial when the whole triad begins rowing together, and the zygians in particular need to be told when they are skying their blades, since it is impossible from that position to gauge correctly blade height independently of coaching.

This initial training can be done with only a third of the ship rowing at a time, or perhaps half the ship; but, if it were possible, it would be better perhaps to have whole levels rowing together, even if the overall timing were somewhat chaotic, so as to maximise water time for as many rowers as possible.

Once a reasonable stroke has been achieved by all three levels, it would be time to have the triads row together. Here, rowing by thirds is particularly useful to allow the thranites to concentrate on synchronisation of the triad without having to worry about the overall timing along the length of the ship. It is particularly important for the blades not to touch one another, and for the thranites to tell the zygians and thalamians what is happening and why, when they do touch. The most vexing thing, especially for a thalamian, is to have constant contact with other blades without knowing where the problem lies and which blade is hitting his and why. The most common problems are varying stroke lengths in the water or out, and differences in timing the catch and finish. Since it is difficult for the thalamians and thranites to hear each other while rowing, the rest period for each third should be used for discussion and analysis of the stroke within each individual triad. The rest periods can be potentially as profitable for building confidence within the triad as the periods of rowing.

The short stroke length in the trieres suggests that ship speed was generated by rowing at quite a high rate of striking, perhaps fifty or more strokes per minute (see the conclusion of Appendix 7). A high rate of striking would compensate for the short stroke of the trieres compared with either sliding-seat racing shells or modern fixed-seat boats. The 1987 crew was capable of rowing at a rate of over forty spm for only a very short time before the weaker members of the crew became incapable of continuing. It seems reasonable to assume that the ancient Athenian crews had no such problem, and though it would be unduly optimistic to assume that a modern crew could ever reach ancient standards in all respects (lack of training time being the obvious obstacle), it is not unduly optimistic to assume that a modern crew should match, if only for a shorter period, the speed of the ancient trieres.

B.9 THE ROWING MASTER'S STORY : Dr Michael Budd

Boris Rankov had been invited to manage the oarcrew and oar trials of the trieres. However, when he slipped a disc in Australia a few days before the start of the trials, I was asked to take over as Rowing Master.

The crew was organised in six teams, three on each side of the ship, made up as far as possible in "buddy groups" and rowing on their preferred side. Each team had an experienced oarsman as leader and another as deputy, to control and coach the rowing under conditions, and in a ship, which was completely strnge to all of us.

At the start of the first outing I had put the most experienced people at the bottom level where they had to row blind (i.e. without seeing the blade of their oar in the water). After a chaotic two hours on the first outing I decided to re-jig the crew completely for the next outing, with tall, experienced people at the top level, short experienced people at the bottom and the less experienced people in the middle level. We then used the thranites (top level oarsmen) to go out and come in to mooring, because they could see what they were doing and keep time, and they could easily hear the words of command. We rowed in sections, using two of the six teams at a time, sometimes using one level at a time, sometimes small vertical sections, gradually increasing the number rowing, but always ending with the entire crew.

After four outings we knew we had a problem. The inexperienced zygians (rowing at middle level), wielding a heavy oar, were in some cases too slow into the water and were crashing down on the thalamians' oars (at the bottom level) below them, driving the oar handle into their throats and trapping their heads against a 4-inch square beam. It was dangerous to say the least.

Peter MacLeod, the Barrack Master, solved the difficulty. ' Get the thranites to put their blades between the zygians' blades and thalamians' blades', he said,' and keep them there'. From then onwards we made rapid progress despite spending half our rowing time making minor repairs to the ship. Loose tholepins and loose nails were the main problem.

The design of the ship was virtually perfect. In the next five days we practised all the manoeuvres we had to carry out for the sea trials, and our Recorders measured and timed all our turns and speed tests. These rapidly improved. Ratings started to climb until we reached 44 strokes per minute. For some of the thalamians this was terrifying, and many of the crew suggested that we should try for a lower, safer rating. After trial and error we found that the ship ran very well at 32-34 spm. with only a slight loss of speed. It was possible for the whole crew to row at 32 for 20 minutes.

Valuable suggestions continued to flow in from the crew. Some were acted on after consultation with the deputy Rowing Masters, Stephen Walters and Tim Watson, and the team leaders. The last week was devoted to the Sea Trials themselves which were completely successful.

On the penultimate day we intended to row round the island, about 12 sea miles. It looked a lot further on the map, a recce making us all apprehensive of the crew's ability to do it. In the event we set off two hours late and rowed to a beach which turned out to be a rubbish heap. Two hours were further wasted mooring and casting off. We also spent time while a television crew took underwater shots of the ship. Then we had a nice sail back to the entrance of Poros bay and a hard row back to our moorings into a headwind. About six miles were covered.

On the morning of the last day we showed off to the town by rowing beautifully up the main shipping channel between Poros island and the mainland of the Peloponnese and back again. In the evening we went out with 60 guests of the Trireme Trust, and gave a demonstration run looking more like a Vietnamese refugee ship than an ancient man-of-war. At the termination of this final trip our Helmsman, Tim Shaw, brought the ship in to her moorings stern first perfectly, winning the congratulations of the Captain.

The crew were, on the whole, magnificent, patient and understanding. Communication down the ship was always difficult. The over-heavy oars were exhausting to handle. There were misunderstandings, largely the result of the language problem, frustrations, and the hindrance of the omnipresent media. These I shall forget, and remember the heat, the talk, the singing in the ship, the feeling of elation when all went well, and the fantastic keenness of many people to make the venture a success; and the wonderful sound of the oars rowing in perfect time.

GLOSSARY

Boat mast/sail	Small mast and sail stepped forward of the main mast (see *AT* p.176-7).
Brace	A rope attached to the end of a yard for swinging the yard so that the sail catches the wind.
Brailing line, brail	A rope attached at the bottom edge of a square sail, led up its forward surface,over the yard and then through a fairlead to a point aft where it is secured. By pulling on the brail the sail area can be reduced, or pulled up completely to the yard.
Butt	Joint between two structural members joined end to end.
Catch	Entry of oarblades into the water at the start of the pulling part of an oar stroke.
Diekplous	The tactical movement of 'breakthough': see General Index.
Displacement	The mass of the water which would be required to fill the immersed volume of the ship: in magnitude it equals the weight of the ship.
Epibatai	Men carried on deck.
Finish	Exit of oarblades at the end of the pulling part of the oar stroke.
Fish	To join two spars of timber together by rope lashings
Free (of wind)	A wind blowing from astern or from either quarter when on a particular course
Gearing (of oars)	The ratio A : B, where A is the distnace from the thole to the centre of pressure of the blade in the water, projected on to a horizontal plane, and B is the horizontal distance of the thole from the plane of symmetry of the oarsman.
GZ	The righting lever separating the buoyancy force from the Centre of Gravity of the ship.
Halyard	A rope for raising the yard from the deck to the top top of the mast.
Heel	Lateral inclination of the ship.
Hog, Hogging	See *AT* p.197-9. Bending or shearing of the hull longitudinally so that the ends drop relative to the middle
Horsepower	1 Horsepower = 746 watts
Hoplite	In ancient Greece, a heavily armed footsoldier.

GLOSSARY

Hypozoma	See General Index and *AT* p.170-172.
Joule	The measure of work : 1 Joule = 1 Newton.Metre
KB	The height of the centre of buoyancy above the underside of the keel (USK). The centre of buoyancy (B) is the centroid (see *AT* Glossary) of the immersed volume of the ship, through which the buoyancy force acts.
KM_T	The height of the transverse metacentre above USK (see KB above). The metacentre is the point in the middle plane of the ship intersected by the line of the buoyancy force when the ship is heeled by a small angle.
LCB	Longitudinal Centre of Buoyancy, the longitudinal position of the Centre of Buoyancy: it is measured from 11 Ordinate (for Ordinate see under LCF).
LCF	Longitudinal Centre of Flotation, the longitudinal position of the centroid of the water plane of the ship: the water plane is the plan shape of the waterline: measured from 11 Ordinate, the mid-length of a particular waterline whose length is divided into 20 equal parts by 21 Ordinates to enable the shape of the ship's hull to be defined.
Metacentric Height (GM)	The height of the metacentre (M) above the centre of gravity (G).
MCT	The moment required to change trim by 1 cm.: trim is the difference between the depths below waterline of the line of the USK (see KB) produced at the forward and after ends of the waterline.
Newton	The measure of force : 1 Newton is the force required to accelerate 1 kilogram mass by one metre per second per second.
Periplous	The tactical movement of 'encirclement': see General Index.
Prismatic Coefficient	A measure of the fullness f the ends of a hull form. It is the ratio of the displacement volume to that of a prism of length equal to the waterline and of section identical with the immersed mid-section of the ship
Quarter	A side of the ship near the stern
Room	Of oars or of oarcrew, the longitudinal separation of tholepins, and therefore also of oarcrew, oars, seats etc.
Sailing Efficiency	The ratio of a ship's speed through the water to the true wind speed
Scarf	A butt joint in which the ends of each member is

shaped to overlap the other to increase its strength

Sheet	A rope holding down the aftermost bottom corner of a square sail
Slip	The longitudinal movement of an oarblade (i.e. movement aft when the oarhandle is being pulled) relative to the water, from the moment of the catch to that of the finish
Spike	Large nail.
Striking, Rate of	Number of oarstrokes performed in one minute (= s.p.m.)
Tabernacle	A strong socket for holding the bottom of the mast, to allow the mast to hinge down from deck level
Tack	A rope holding down the forward bottom corner of a square sail
Tarros	Greek word for the complete oarsystem of an oared warship
Thalamian	English term from Greek *thalamios*, the name for the oarsmen in the lowest file of the trieres
Thole, tholepin	The pin of wood forming, with the oarloop, the fulcrum of the oar
Thranite	English term from Greek *thranites*, the name for the oarsmen in the uppermost file of the trieres
Tonnes per cm. immersion (TPC)	The amount in tonnes by which the displacement changes per 1 cm rise or sinkage of the ship in the water
Triad	Of oarcrew, three rowers sitting roughly in the same vertical plane of the trieres on one side, whose oarwork is controlled by the uppermost member of the triad, the thranite
Trierarch	The captain of a trieres
Wale	An assemblage of thick and broad planks on the outside of the ship's hull
Watt	The measure of power: 1 watt = 1 Joule/sec.
Yard	Timber spar, hung transversely across a mast to spread a square sail
Zygian	English word from Greek *zygios*, the name for the oarsmen in the (vertically) middle file of the trieres.

BIBLIOGRAPHY

Basch, L. (1987) *Le Musée imaginaire de la marine antique*. Athens = ***MIMA***
Blackman, D.J. (1968) 'The Ship-sheds' Chapter 8 of *GOS*
Bourne, Gilbert C. (1925) *A textbook of oarsmanship*. Oxford

Coates, J.F. *Specifications [of the trieres]* English edition : unpublished.
Coates, J.F. (1987) 'The strength and behaviour of tension-tourniquets, or Spanish windlasses, of natural fibre ropes' *IJNA* 16.3: p.207-211
Coates, J.F. and Sean McGrail edd. (1984) *The Greek Trireme of the 5th century BC* .London
Coates, John and Morrison, John (1987) 'Authenticity in the replica Athenian trieres' *Antiquity* 61.231: p.87-90.
Cook, Theodore (1909) *Rowing at Henley*. Oxford
Creuze, A.F.B. (1841) *Treatise on Naval Architecture*. Edinburgh

Dragatzes, I.X. (1885) 'Report on the excavations at Piraeus' *Praktika*

Edwards, H.R.A. (1963) *The way of a man with a blade*. London

Frost, H. (1974) 'The Punic wreck in Sicily' *IJNA* 3.1: p.35-52

Hammond, N.G.L. (1962) 'The narrative of Herodotus VII and the Decree of Themistocles at Troezen' *JHS* 102: p.75-93
Humphreys, S.C. (1978) 'Artists' Mistakes' *IJNA* 7.1: p.78

Landels, J.G.(1980) *Engineering in the Ancient World*. London
Lazenby, J.F.(1987)'The Diekplous' *Greece and Rome* 2nd series 34: p.169-77
Lehmann, R.C.(1898) *Rowing* Isthmian Library. London
Loukakis, T. (1985) *Report NAL 06-F-1985* Technical University, Athens

McGrail,Sean and Farrell,Anthony (1979) 'Rowing: aspects of the ethnographic and iconographic evidence' *IJNA 8.2* : pp.156-166
Meiggs, Russell and Lewis, David (1969) *Greek Historical Inscriptions to the end of the 5th century BC*. Oxford = Meiggs-Lewis
Morrison, J.S. and Coates, J.F.(1986) *The Athenian Trireme*. Cambridge = *AT*
Morrison, J.S. and Williams,R.T. (1968) *Greek Oared Ships*. Cambridge = *GOS*

Roberts, O.T.P. (1987) 'The Rig of *Olympias*' *Proceedings of the Second Symposium on Ship Construction in the Ancient World, Delphi*

Steffy, J.R. (1983)' The Athlit Ram ' ***MM*** 69.3: p.259-269

Tarn, W.W. (1905) 'The Greek Warship' *JHS* 25: pp.137-73, 204-224
Tarn, W.W. (1938) 'The Oarage of Greek Warships' ***MM*** 19: p.52-74

Welsh,Frank (1958) *Building the Trireme* London
Whitehead, I.M.(1987)' The Periplous' *Greece and Rome* 2nd ser. 34: pp.178-85
Woodgate, W.B.(1889) *Boating* 2nd Edition Badminton Library. London

INDEX OF ANCIENT AUTHORS CITED

INDEX OF ANCIENT AUTHORS CITED

GENERAL INDEX

www.ingramcontent.com/pod-product-compliance
Lightning Source LLC
Chambersburg PA
CBHW040103290626
47361CB00023B/1266